These safety symbols are used in laboratory and field investigations... ...ing of each symbol and refer to this page often. *Remember to was...*

PROTECTIVE EQUIPMENT Do not begin any lab withou...

 GOGGLES Proper eye protection must be worn when performing or observing science activities that involve items or conditions as listed below.

 APRON Wear an approved apron when using substances that could stain, wet, or destroy cloth.

removing goggles and after all lab activities.

materials, chemicals, animals, or materials that can stain or irritate hands.

LABORATORY HAZARDS

Symbols	Potential Hazards	Precaution	Response
DISPOSAL	contamination of classroom or environment due to improper disposal of materials such as chemicals and live specimens	• DO NOT dispose of hazardous materials in the sink or trash can. • Dispose of wastes as directed by your teacher.	• If hazardous materials are disposed of improperly, notify your teacher immediately.
EXTREME TEMPERATURE	skin burns due to extremely hot or cold materials such as hot glass, liquids, or metals; liquid nitrogen; dry ice	• Use proper protective equipment, such as hot mitts and/or tongs, when handling objects with extreme temperatures.	• If injury occurs, notify your teacher immediately.
SHARP OBJECTS	punctures or cuts from sharp objects such as razor blades, pins, scalpels, and broken glass	• Handle glassware carefully to avoid breakage. • Walk with sharp objects pointed downward, away from you and others.	• If broken glass or injury occurs, notify your teacher immediately.
ELECTRICAL	electric shock or skin burn due to improper grounding, short circuits, liquid spills, or exposed wires	• Check condition of wires and apparatus for fraying or uninsulated wires, and broken or cracked equipment. • Use only GFCI-protected outlets	• DO NOT attempt to fix electrical problems. Notify your teacher immediately.
CHEMICAL	skin irritation or burns, breathing difficulty, and/or poisoning due to touching, swallowing, or inhalation of chemicals such as acids, bases, bleach, metal compounds, iodine, poinsettias, pollen, ammonia, acetone, nail polish remover, heated chemicals, mothballs, and any other chemicals labeled or known to be dangerous	• Wear proper protective equipment such as goggles, apron, and gloves when using chemicals. • Ensure proper room ventilation or use a fume hood when using materials that produce fumes. • NEVER smell fumes directly. • NEVER taste or eat any material in the laboratory.	• If contact occurs, immediately flush affected area with water and notify your teacher. • If a spill occurs, leave the area immediately and notify your teacher.
FLAMMABLE	unexpected fire due to liquids or gases that ignite easily such as rubbing alcohol	• Avoid open flames, sparks, or heat when flammable liquids are present.	• If a fire occurs, leave the area immediately and notify your teacher.
OPEN FLAME	burns or fire due to open flame from matches, Bunsen burners, or burning materials	• Tie back loose hair and clothing. • Keep flame away from all materials. • Follow teacher instructions when lighting and extinguishing flames. • Use proper protection, such as hot mitts or tongs, when handling hot objects.	• If a fire occurs, leave the area immediately and notify your teacher.
ANIMAL SAFETY	injury to or from laboratory animals	• Wear proper protective equipment such as gloves, apron, and goggles when working with animals. • Wash hands after handling animals.	• If injury occurs, notify your teacher immediately.
BIOLOGICAL	infection or adverse reaction due to contact with organisms such as bacteria, fungi, and biological materials such as blood, animal or plant materials	• Wear proper protective equipment such as gloves, goggles, and apron when working with biological materials. • Avoid skin contact with an organism or any part of the organism. • Wash hands after handling organisms.	• If contact occurs, wash the affected area and notify your teacher immediately.
FUME	breathing difficulties from inhalation of fumes from substances such as ammonia, acetone, nail polish remover, heated chemicals, and mothballs	• Wear goggles, apron, and gloves. • Ensure proper room ventilation or use a fume hood when using substances that produce fumes. • NEVER smell fumes directly.	• If a spill occurs, leave area and notify your teacher immediately.
IRRITANT	irritation of skin, mucous membranes, or respiratory tract due to materials such as acids, bases, bleach, pollen, mothballs, steel wool, and potassium permanganate	• Wear goggles, apron, and gloves. • Wear a dust mask to protect against fine particles.	• If skin contact occurs, immediately flush the affected area with water and notify your teacher.
RADIOACTIVE	excessive exposure from alpha, beta, and gamma particles	• Remove gloves and wash hands with soap and water before removing remainder of protective equipment.	• If cracks or holes are found in the container, notify your teacher immediately.

Your online portal to everything you need

connectED.mcgraw-hill.com

Look for these icons to access exciting digital resources

 Video

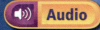 Audio

 Review

Inquiry

WebQuest

Assessment

Concepts in Motion

WATER AND OTHER RESOURCES

Glencoe

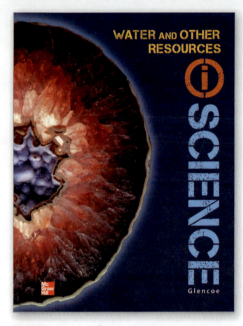

Geode

This is a cross section of geode, a type of rock. The outside of a geode is generally limestone, but the inside contains mineral crystals. Crystals only partially fill this geode, but other geodes are filled completely with crystals.

 Education

Copyright © 2012 The McGraw-Hill Companies, Inc. All rights reserved. No part of this publication may be reproduced or distributed in any form or by any means, or stored in a database or retrieval system, without the prior written consent of The McGraw-Hill Companies, Inc., including, but not limited to, network storage or transmission, or broadcast for distance learning.

Send all inquiries to:
McGraw-Hill Education
8787 Orion Place
Columbus, OH 43240-4027

ISBN: 978-0-07-888011-7
MHID: 0-07-888011-4

Printed in the United States of America.

2 3 4 5 6 7 8 9 10 11   DOW   15 14 13 12 11

Authors and Contributors

Authors

American Museum of Natural History
New York, NY

Michelle Anderson, MS
Lecturer
The Ohio State University
Columbus, OH

Juli Berwald, PhD
Science Writer
Austin, TX

John F. Bolzan, PhD
Science Writer
Columbus, OH

Rachel Clark, MS
Science Writer
Moscow, ID

Patricia Craig, MS
Science Writer
Bozeman, MT

Randall Frost, PhD
Science Writer
Pleasanton, CA

Lisa S. Gardiner, PhD
Science Writer
Denver, CO

Jennifer Gonya, PhD
The Ohio State University
Columbus, OH

Mary Ann Grobbel, MD
Science Writer
Grand Rapids, MI

Whitney Crispen Hagins, MA, MAT
Biology Teacher
Lexington High School
Lexington, MA

Carole Holmberg, BS
Planetarium Director
Calusa Nature Center and Planetarium, Inc.
Fort Myers, FL

Tina C. Hopper
Science Writer
Rockwall, TX

Jonathan D. W. Kahl, PhD
Professor of Atmospheric Science
University of Wisconsin-Milwaukee
Milwaukee, WI

Nanette Kalis
Science Writer
Athens, OH

S. Page Keeley, MEd
Maine Mathematics and Science Alliance
Augusta, ME

Cindy Klevickis, PhD
Professor of Integrated Science and Technology
James Madison University
Harrisonburg, VA

Kimberly Fekany Lee, PhD
Science Writer
La Grange, IL

Michael Manga, PhD
Professor
University of California, Berkeley
Berkeley, CA

Devi Ried Mathieu
Science Writer
Sebastopol, CA

Elizabeth A. Nagy-Shadman, PhD
Geology Professor
Pasadena City College
Pasadena, CA

William D. Rogers, DA
Professor of Biology
Ball State University
Muncie, IN

Donna L. Ross, PhD
Associate Professor
San Diego State University
San Diego, CA

Marion B. Sewer, PhD
Assistant Professor
School of Biology
Georgia Institute of Technology
Atlanta, GA

Julia Meyer Sheets, PhD
Lecturer
School of Earth Sciences
The Ohio State University
Columbus, OH

Michael J. Singer, PhD
Professor of Soil Science
Department of Land, Air and Water Resources
University of California
Davis, CA

Karen S. Sottosanti, MA
Science Writer
Pickerington, Ohio

Paul K. Strode, PhD
I.B. Biology Teacher
Fairview High School
Boulder, CO

Jan M. Vermilye, PhD
Research Geologist
Seismo-Tectonic Reservoir Monitoring (STRM)
Boulder, CO

Judith A. Yero, MA
Director
Teacher's Mind Resources
Hamilton, MT

Dinah Zike, MEd
Author, Consultant,
Inventor of Foldables
Dinah Zike Academy;
Dinah-Might Adventures, LP
San Antonio, TX

Margaret Zorn, MS
Science Writer
Yorktown, VA

iii

Consulting Authors

Alton L. Biggs
Biggs Educational Consulting
Commerce, TX

Ralph M. Feather, Jr., PhD
Assistant Professor
Department of Educational
Studies and Secondary
Education
Bloomsburg University
Bloomsburg, PA

Douglas Fisher, PhD
Professor of Teacher Education
San Diego State University
San Diego, CA

Edward P. Ortleb
Science/Safety Consultant
St. Louis, MO

Series Consultants

Science

Solomon Bililign, PhD
Professor
Department of Physics
North Carolina Agricultural
and Technical State University
Greensboro, NC

John Choinski
Professor
Department of Biology
University of Central Arkansas
Conway, AR

Anastasia Chopelas, PhD
Research Professor
Department of Earth and
Space Sciences
UCLA
Los Angeles, CA

David T. Crowther, PhD
Professor of Science Education
University of Nevada, Reno
Reno, NV

A. John Gatz
Professor of Zoology
Ohio Wesleyan University
Delaware, OH

Sarah Gille, PhD
Professor
University of California
San Diego
La Jolla, CA

David G. Haase, PhD
Professor of Physics
North Carolina State
University
Raleigh, NC

Janet S. Herman, PhD
Professor
Department of Environmental
Sciences
University of Virginia
Charlottesville, VA

David T. Ho, PhD
Associate Professor
Department of Oceanography
University of Hawaii
Honolulu, HI

Ruth Howes, PhD
Professor of Physics
Marquette University
Milwaukee, WI

Jose Miguel Hurtado, Jr., PhD
Associate Professor
Department of Geological
Sciences
University of Texas at El Paso
El Paso, TX

Monika Kress, PhD
Assistant Professor
San Jose State University
San Jose, CA

Mark E. Lee, PhD
Associate Chair & Assistant
Professor
Department of Biology
Spelman College
Atlanta, GA

Linda Lundgren
Science writer
Lakewood, CO

Series Consultants, continued

Keith O. Mann, PhD
Ohio Wesleyan University
Delaware, OH

Charles W. McLaughlin, PhD
Adjunct Professor of Chemistry
Montana State University
Bozeman, MT

Katharina Pahnke, PhD
Research Professor
Department of Geology and Geophysics
University of Hawaii
Honolulu, HI

Jesús Pando, PhD
Associate Professor
DePaul University
Chicago, IL

Hay-Oak Park, PhD
Associate Professor
Department of Molecular Genetics
Ohio State University
Columbus, OH

David A. Rubin, PhD
Associate Professor of Physiology
School of Biological Sciences
Illinois State University
Normal, IL

Toni D. Sauncy
Assistant Professor of Physics
Department of Physics
Angelo State University
San Angelo, TX

Malathi Srivatsan, PhD
Associate Professor of Neurobiology
College of Sciences and Mathematics
Arkansas State University
Jonesboro, AR

Cheryl Wistrom, PhD
Associate Professor of Chemistry
Saint Joseph's College
Rensselaer, IN

Reading

ReLeah Cossett Lent
Author/Educational Consultant
Blue Ridge, GA

Math

Vik Hovsepian
Professor of Mathematics
Rio Hondo College
Whittier, CA

Series Reviewers

Thad Boggs
Mandarin High School
Jacksonville, FL

Catherine Butcher
Webster Junior High School
Minden, LA

Erin Darichuk
West Frederick Middle School
Frederick, MD

Joanne Hedrick Davis
Murphy High School
Murphy, NC

Anthony J. DiSipio, Jr.
Octorara Middle School
Atglen, PA

Adrienne Elder
Tulsa Public Schools
Tulsa, OK

Series Reviewers, continued

Carolyn Elliott
Iredell-Statesville Schools
Statesville, NC

Christine M. Jacobs
Ranger Middle School
Murphy, NC

Jason O. L. Johnson
Thurmont Middle School
Thurmont, MD

Felecia Joiner
Stony Point Ninth Grade Center
Round Rock, TX

Joseph L. Kowalski, MS
Lamar Academy
McAllen, TX

Brian McClain
Amos P. Godby High School
Tallahassee, FL

Von W. Mosser
Thurmont Middle School
Thurmont, MD

Ashlea Peterson
Heritage Intermediate Grade Center
Coweta, OK

Nicole Lenihan Rhoades
Walkersville Middle School
Walkersvillle, MD

Maria A. Rozenberg
Indian Ridge Middle School
Davie, FL

Barb Seymour
Westridge Middle School
Overland Park, KS

Ginger Shirley
Our Lady of Providence Junior-Senior High School
Clarksville, IN

Curtis Smith
Elmwood Middle School
Rogers, AR

Sheila Smith
Jackson Public School
Jackson, MS

Sabra Soileau
Moss Bluff Middle School
Lake Charles, LA

Tony Spoores
Switzerland County Middle School
Vevay, IN

Nancy A. Stearns
Switzerland County Middle School
Vevay, IN

Kari Vogel
Princeton Middle School
Princeton, MN

Alison Welch
Wm. D. Slider Middle School
El Paso, TX

Linda Workman
Parkway Northeast Middle School
Creve Coeur, MO

Teacher Advisory Board

The Teacher Advisory Board gave the authors, editorial staff, and design team feedback on the content and design of the Student Edition. They provided valuable input in the development of *Glencoe ⓘScience*.

Frances J. Baldridge
Department Chair
Ferguson Middle School
Beavercreek, OH

Jane E. M. Buckingham
Teacher
Crispus Attucks Medical
Magnet High School
Indianapolis, IN

Elizabeth Falls
Teacher
Blalack Middle School
Carrollton, TX

Nelson Farrier
Teacher
Hamlin Middle School
Springfield, OR

Michelle R. Foster
Department Chair
Wayland Union
Middle School
Wayland, MI

Rebecca Goodell
Teacher
Reedy Creek Middle School
Cary, NC

Mary Gromko
Science Supervisor K–12
Colorado Springs District 11
Colorado Springs, CO

Randy Mousley
Department Chair
Dean Ray Stucky
Middle School
Wichita, KS

David Rodriguez
Teacher
Swift Creek Middle School
Tallahassee, FL

Derek Shook
Teacher
Floyd Middle Magnet School
Montgomery, AL

Karen Stratton
Science Coordinator
Lexington School District One
Lexington, SC

Stephanie Wood
Science Curriculum Specialist,
K–12
Granite School District
Salt Lake City, UT

Online Guide

connectED.mcgraw-hill.com

▶ **Your Digital Science Portal**

See the science in real life through these exciting videos.

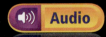

Click the link and you can listen to the text while you follow along.

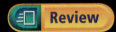

Try these interactive tools to help you review the lesson concepts.

Explore concepts through hands-on and virtual labs.

These web-based challenges relate the concepts you're learning about to the latest news and research.

Digital and Print Solutions

The icons in your online student edition link you to interactive learning opportunities. Browse your online student book to find more.

Review — Personal Tutor

Concepts in Motion — Animation

"It's **easy** to do my assignments online and **quick** to find everything I need."

Assessment
Check how well you understand the concepts with online quizzes and practice questions.

Concepts in Motion
The textbook comes alive with animated explanations of important concepts.

Multilingual eGlossary
Read key vocabulary in 13 languages.

Treasure Hunt

Your science book has many features that will aid you in your learning. Some of these features are listed below. You can use the activity at the right to help you find these and other special features in the book.

- **THE BIG IDEA** can be found at the start of each chapter.
- The Reading Guide at the start of each lesson lists **Key Concepts**, vocabulary terms, and online supplements to the content.
- **ConnectED** icons direct you to online resources such as animations, personal tutors, math practices, and quizzes.
- **Inquiry** Labs and Skill Practices are in each chapter.
- Your **FOLDABLES** help organize your notes.

START

1. What four margin items can help you build your vocabulary?

2. On what page does the glossary begin? What glossary is online?

3. In which Student Resource at the back of your book can you find a listing of Laboratory Safety Symbols?

4. Suppose you want to find a list of all the Launch Labs, MiniLabs, Skill Practices, and Labs, where do you look?

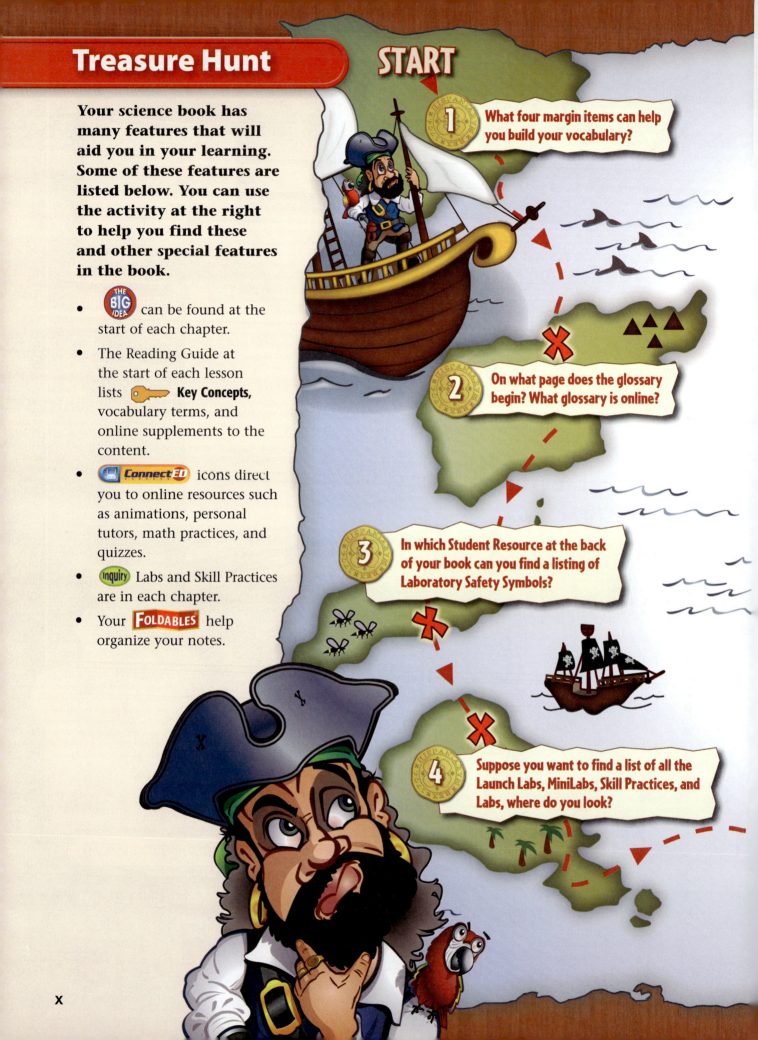

Table of Contents

Unit 4 **Water and Other Resources** **520**

Chapter 15 Earth's Water **524**
- **Lesson 1** The Water Planet...526
- **Lesson 2** The Properties of Water 536
  - **Skill Practice** Why is liquid water denser than ice?.................544
- **Lesson 3** Water Quality 545
  - **Lab** Temperature and Water's Density..................................552

Chapter 16 Oceans **560**
- **Lesson 1** Composition and Structure of Earth's Oceans 562
- **Lesson 2** Ocean Waves and Tides 572
  - **Skill Practice** High Tides in the Bay of Fundy 579
- **Lesson 3** Ocean Currents 580
  - **Skill Practice** How do oceanographers study ocean currents? 587
- **Lesson 4** Environmental Impacts on Oceans 588
  - **Lab** Predicting Whale Sightings Based on Upwelling 596

Chapter 17 Freshwater **604**
- **Lesson 1** Glaciers and Polar Ice Sheets 606
- **Lesson 2** Streams and Lakes 616
  - **Skill Practice** How does water flow into and out of streams? 623
- **Lesson 3** Groundwater and Wetlands 624
  - **Lab** What can be done about pollution? 632

Chapter 18 Natural Resources **640**
- **Lesson 1** Energy Resources 642
  - **Skill Practice** How can you identify bias and its source? 651
- **Lesson 2** Renewable Energy Resources.....................................652
  - **Skill Practice** How can you analyze energy-use data for information to help conserve energy?..659
- **Lesson 3** Land Resources 660
- **Lesson 4** Air and Water Resources..668
  - **Lab** Research Efficient Energy and Resource Use 674

Table of Contents

Student Resources

Science Skill Handbook ... SR-2
Scientific Methods ... SR-2
Safety Symbols ... SR-11
Safety in the Science Laboratory ... SR-12

Math Skill Handbook ... SR-14
Math Review ... SR-14
Science Application ... SR-24

Foldables Handbook ... SR-29

Reference Handbook ... SR-40
Periodic Table of the Elements ... SR-40
Topographic Map Symbols ... SR-42
Rocks ... SR-43
Minerals ... SR-44
Weather Map Symbols ... SR-46

Glossary ... G-2
Index ... I-2
Credits ... C-2

Inquiry

Inquiry Launch Labs

15-1	Which heats faster?	527
15-2	How many drops can fit on a penny?	537
15-3	How can you test the cloudiness of water?	546
16-1	How are salt and density related?	563
16-2	How is sea level measured?	573
16-3	How does wind move water?	581
16-4	What happens to litter in the oceans?	589
17-1	Where is all the water on Earth?	607
17-2	How can you measure the health of a stream?	617
17-3	How solid is Earth's surface?	625
18-1	How do you use energy resources?	643
18-2	How can renewable energy sources generate energy in your home?	653
18-3	What resources from the land do you use every day?	661
18-4	How often do you use water each day?	669

Inquiry MiniLabs

15-1	What happens to temperature during change of state?	531
15-2	Is every substance less dense in its solid state?	541
15-3	How do oxygen levels affect marine life?	548
16-1	How does salinity affect the density of water?	565
16-2	Can you analyze tidal data?	577
16-3	How does temperature affect ocean currents?	585
16-4	How does the pH of seawater affect marine organisms?	594
17-1	Does the ground's color affect temperature?	611
17-2	How does a thermocline affect pollution in a lake?	620
17-3	Can you model freshwater environments?	630
18-1	What is your reaction?	647
18-2	How are renewable energy resources used at your school?	657
18-3	How can you manage land resource use with environmental responsibility?	662
18-4	How much water can a leaky faucet waste?	671

Inquiry

Inquiry Skill Practice

15-2 Why is liquid water denser than ice? 544
16-2 High Tides in the Bay of Fundy 579
16-3 How do oceanographers study ocean currents? 587
17-2 How does water flow into and out of streams? 623
18-1 How can you identify bias and its source? 651
18-2 How can you analyze energy-use data for information to help conserve energy? 659

Inquiry Labs

15-3 Temperature and Water's Density 552
16-4 Predicting Whale Sightings Based on Upwelling 596
17-3 What can be done about pollution? 632
18-4 Research Efficient Energy and Resource Use 674

Features

GREEN SCIENCE

18-3 A Greener Greensburg 667

SCIENCE & SOCIETY

17-1 Life at the Top of the World 615

CAREERS in SCIENCE

15-1 Oceans on the Rise—Again 535
16-1 Exploring Deep-Sea Vents 571

xv

Unit 4
WATER AND OTHER RESOURCES

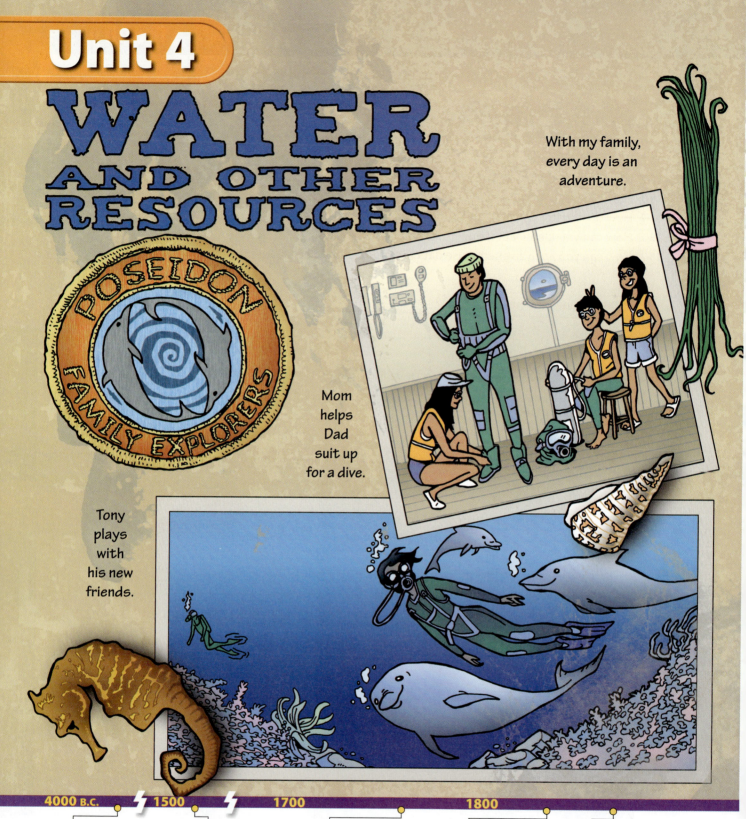

With my family, every day is an adventure.

Mom helps Dad suit up for a dive.

Tony plays with his new friends.

4000 B.C.

3500 B.C.
Egyptians develop and craft sailing vessels, most likely to use in the eastern Mediterranean, near the mouth of the Nile River.

1500

1519–1522
Ferdinand Magellan's crew attempts to circumnavigate the world via ship; one ship succeeds.

1700

1768–1780
James Cook explores the southern parts of the oceans looking for Antarctica. He is the first to use a chronometer, a precise clock, to determine longitude.

1800

1831–1836
Charles Darwin sails on the H.M.S. *Beagle* to the Galápagos Islands. His research there leads him to develop the concept of natural selection.

1872–1876
The H.M.S. *Challenger* travels around the world collecting sediment and water samples, soundings, and biological specimens.

1900	1950			2000

1925–1927 A German vessel, *The Meteor,* sails the Atlantic taking sonar measurements. During this expedition the Mid-Atlantic Ridge is discovered.

1947 Satellite data leads to worldwide mapping of seafloor geography from space.

1958 The nuclear submarine *Nautilus* makes the first undersea voyage to the North Pole.

1960 The deep-sea submersible vessel *Trieste* reaches a depth of 10,915 m at the Challenger Deep in the Mariana Trench. This location is believed to be the deepest part of any ocean on Earth.

Inquiry Visit ConnectED for this unit's STEM activity.

Charts, Tables, and Graphs

Imagine that 3 seconds are left in the semifinal game of your favorite sporting event. The clock runs out, and the buzzer sounds! You cheer as your team advances to the finals! You grab the bracket that you made and record another win.

A bracket organizes and displays the wins and losses of teams in a tournament, as shown in **Figure 1.** Brackets, like maps, tables, and graphs, are a type of chart. A **chart** is a visual display that organizes information. Charts help you organize data. Charts also help you identify patterns, trends, or errors in your data and communicate data to others.

What are tables?

Suppose you volunteer for a cleanup program at a local beach. The organizers need to know the types of debris found at different times of the year. Each month, you collect debris, separate it into categories, and weigh each category of debris. You record your data in a table. A **table** is a type of chart that organizes related data in columns and rows. Titles are usually placed at the top of each column or at the beginning of each row to help organize the data, as shown in **Table 1.**

▲ **Figure 1** A sporting bracket is a type of chart that easily enables you to see which team has won the most games in a tournament.

What are graphs?

A table contains data but it does not clearly show relationships among data. However, displaying data as a graph does clearly show relationships. A **graph** is a type of chart that shows relationships between variables. The organizers of the cleanup program could make different types of graphs from the information in your table to help them better analyze the data.

Table 1 This table organizes data on collected debris into rows and columns so measurements can easily be recorded, compared, and used. ▼

Table 1 Types and Amounts of Debris							
Types of Debris	Jan	Mar	May	July	Sept	Nov	Total for Year
Plastic	3.0	3.5	3.8	4.0	3.7	3.0	21.0
Polystyrene	0.5	1.3	3.2	4.0	2.5	1.2	12.7
Glass	0.8	1.2	1.5	2.0	1.5	1.0	8.0
Rubber	1.1	1.0	1.3	1.5	1.2	1.3	7.4
Metal	1.0	1.0	1.1	1.4	1.1	1.0	6.6
Paper	1.3	1.1	1.5	1.5	0.8	0.3	6.5
Total for Month	9.4	10.6	13.1	15.1	12.1	9.5	69.8

Circle Graphs

If the cleanup organizers want to know the most common type of debris, they will probably use a circle graph. A circle graph shows the percentage of the total that each category represents. This circle graph shows that plastic makes up the largest percentage of debris. The cleanup organizers could then place plastic recycling barrels on the beach so people can recycle their plastic trash.

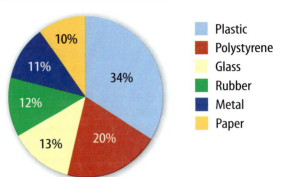

Beach Debris Distribution

Line Graphs

Suppose the cleanup organizers want to know how the total amount of debris on the beach changes throughout the year. They probably will use a line graph. This line graph shows that volunteers collected more debris in summer than in winter. The cleanup organizers could then create a public service announcement for radio stations that reminds beachgoers to throw trash into trash cans and recycling barrels while visiting the beach.

Bar Graphs

Volunteers collected the most debris in July. The cleanup organizers want to know how much of each type of debris volunteers collected in July. Bar graphs are useful for comparing different categories of measurements. This bar graph shows that 4 kg of both plastic and polystyrene were collected in July. The cleanup organizers could then suggest that beach concession stands use smaller, recyclable food containers.

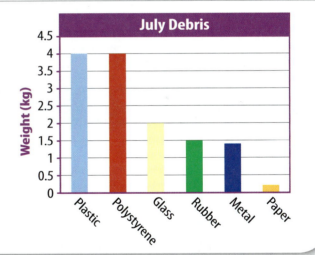

Inquiry MiniLab   25 minutes

How can graphs keep the beach clean?

Suppose you work with the cleanup organizers. What information do you need to make recommendations for keeping the beach cleaner?

1. Based on the type of information in **Table 1,** write a new question about beach debris.
2. Make a graph that allows you to answer your question.

Analyze and Conclude

1. **Distinguish** How did you decide what type of graph to make?
2. **Explain** How did you use your graph to answer your question?
3. **Modify** What recommendations can you make based on your analysis of your graph?

Chapter 15

Earth's Water

THE BIG IDEA What role does water play on Earth?

Inquiry Why are they there?

Animals that live on the dry grasslands of Africa might travel great distances to find water. All living things need water to survive.

- Why is water so important to the animals?
- How did the water get there?
- What role does water play on Earth?

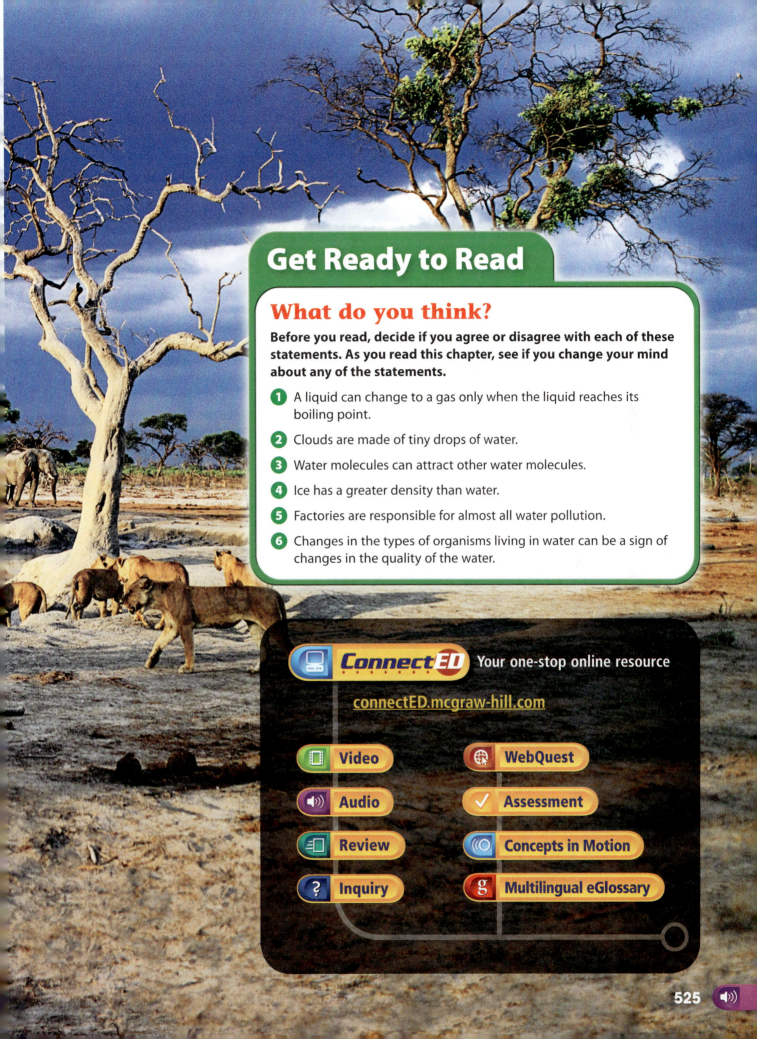

Get Ready to Read

What do you think?
Before you read, decide if you agree or disagree with each of these statements. As you read this chapter, see if you change your mind about any of the statements.

1. A liquid can change to a gas only when the liquid reaches its boiling point.
2. Clouds are made of tiny drops of water.
3. Water molecules can attract other water molecules.
4. Ice has a greater density than water.
5. Factories are responsible for almost all water pollution.
6. Changes in the types of organisms living in water can be a sign of changes in the quality of the water.

Lesson 1

Reading Guide

Key Concepts 🔑
ESSENTIAL QUESTIONS

- Why is water important to life?
- How is water distributed on Earth?
- How is water cycled on Earth?

Vocabulary
specific heat p. 529
hydrosphere p. 530
evaporation p. 531
condensation p. 531
water cycle p. 532
transpiration p. 533

Multilingual eGlossary

Video BrainPOP®

The Water Planet

Inquiry A Water Home?

Water is home to these fish. Like all life on Earth, fish need water to survive, but they also depend on water for a habitat. Where does all the water come from?

Inquiry Launch Lab

20 minutes

Which heats faster?

Water and land heat and cool at different rates. This difference in heating and cooling influences climate.

1. Read and complete a lab safety form.
2. Place two **pie pans** on a flat surface. Fill one with **water** and the other with **soil**.
3. Use **thermometers** to measure the temperature of both materials. Record your measurements in your Science Journal.
4. Place a **lamp** over the pans. Turn on the lamp, and measure the temperature of the water and soil every 5 minutes for 15 minutes.

Think About This

1. Compare the rates at which the two materials heated.
2. 🔑 **Key Concept** Imagine visiting the ocean in summer. Would you expect the climate near the ocean to be warmer or cooler than the climate inland? Why?

Why is water important to life?

You might have read news headlines such as "NASA Searches for Water on Mars" or "Water Found on Saturn's Moon!" Have you ever wondered why scientists are always looking for water in other areas of our solar system? Water is necessary for life. Scientists look for water in other areas of the solar system as a first step to finding life in these areas.

Water is extremely important on Earth for other reasons. Earth's climate is influenced by ocean currents that move thermal energy, commonly called heat, around Earth. Large bodies of water affect local weather patterns as well. Many organisms, such as the jellyfish in **Figure 1**, have water habitats. People also use water for transporting goods and for recreation.

Biological Functions

Water is necessary for the life processes of all living organisms, from a unicellular bacterium to a blue whale. Did you know that the body of a jellyfish is about 95 percent water? Also, about 60 percent of the mass of the human body is made up of water. Even plant seeds that seem dry have a small amount of water inside them.

Figure 1 Water is the habitat of these jellyfish, but all organisms on Earth depend on water for life.

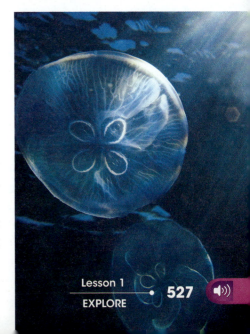

Transport One of the main roles of water in an organism is to transport materials. Water carries nutrients, such as proteins, to cells and even within the cells. It also carries wastes away from cells.

Photosynthesis Water is essential for chemical reactions, such as photosynthesis, to occur within living things. Recall that during photosynthesis, carbon dioxide and water, in the presence of light, react and produce sugar and oxygen. Photosynthesis occurs in plants, algae, and some bacteria. Organisms that undergo photosynthesis are the beginning of almost every food chain.

Body Temperature Regulation Water is an important factor in preventing an organism's body temperature from becoming too high or too low. In humans, as water from the skin, or sweat, changes to a gas, thermal energy transfers to the surrounding air. This helps keep the body cool.

Warming Earth

One reason life can exist on Earth is that Earth's atmosphere traps thermal energy from the Sun. This process is called the greenhouse effect. Some of the Sun's energy that reaches Earth's surface is absorbed and then emitted back toward space. Gases in Earth's atmosphere, such as water vapor (H_2O), methane (CH_4), and carbon dioxide (CO_2), absorb some of this energy and emit it back toward Earth, as shown in **Figure 2.**

Of all the greenhouse gases in the atmosphere, the concentration of water vapor is the highest. Without the greenhouse effect, Earth's average surface temperature would be about −18°C. All the water at Earth's surface would be ice and no organisms could survive at that temperature.

 Reading Check Explain how water helps to heat Earth.

The Greenhouse Effect

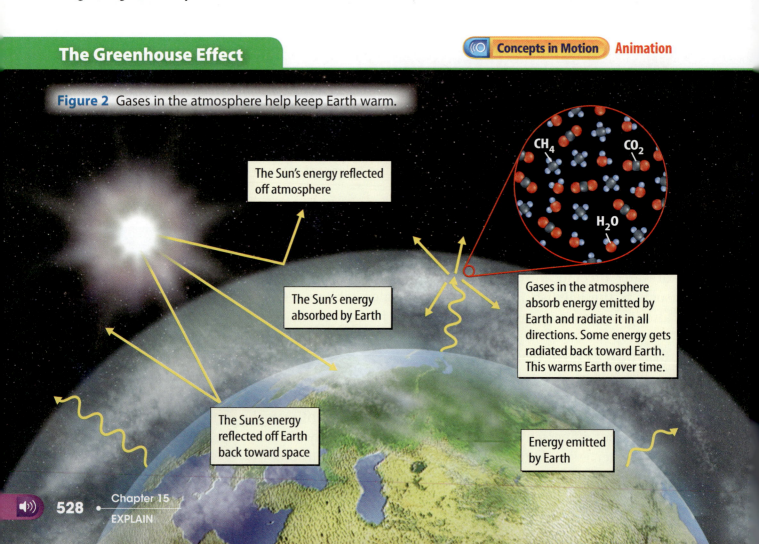

Figure 2 Gases in the atmosphere help keep Earth warm.

Keeping Earth's Temperature Stable

Think about what happens at the beach on a hot day. If you walk barefoot across the sand, you might burn the bottoms of your feet. But when you reach the water, it is refreshingly cool. Why does the water have a lower temperature than the sand?

Water has a high specific heat. **Specific heat** *is the amount of thermal energy needed to raise the temperature of 1 kg of a material by 1°C.* The specific heat of water is about six times higher than the specific heat of sand. That means the water would have to absorb six times as much thermal energy in order to have the same temperature as the sand.

Water's high specific heat is important to life on Earth for several reasons. Water vapor in the air helps control the rate at which air temperature changes. The temperature of water vapor changes slowly. As a result, the temperature change from one season to the next is gradual. Large bodies of water, such as oceans, also heat and cool slowly. This provides a more stable temperature for aquatic organisms and affects climate in coastal areas. The local weather patterns of inland areas near large lakes are affected as well. Examples of how water is important to life are summarized in **Table 1**.

 Key Concept Check Why is water important to life?

Math Skills

Use Equations

To calculate the energy needed to change an object's temperature, use the following equation:

Energy = **specific heat × mass × change in temperature** or,

J = **J/kg · °C × kg × °C**

To solve this equation, you need the object's specific heat. For example, how much energy will raise the temperature of **2 kg** of iron from **20°C** to **30°C**? The specific heat of iron is **460 kg · °C**.

J = **460 kg · °C × 2 kg × 10°C**

The amount of energy is 9,200 J.

Practice

If the specific heat of aluminum is 900 J/kg · °C, how much energy is needed to raise the temperature of a 3 kg sample from 35°C to 45°C?

- Math Practice
- Personal Tutor

Table 1 Importance of Water to Life on Earth

Importance to Life	Examples	
Biological functions	• transport of nutrients and wastes to and from cells • photosynthesis • body temperature regulation	
Keeps Earth warm	• greenhouse effect • air temperature regulation	
Stabilizes Earth's temperature	• gradual temperature change from one season to the next • high specific heat causes large bodies of water to heat up and cool down slowly • stable temperature for aquatic organisms	

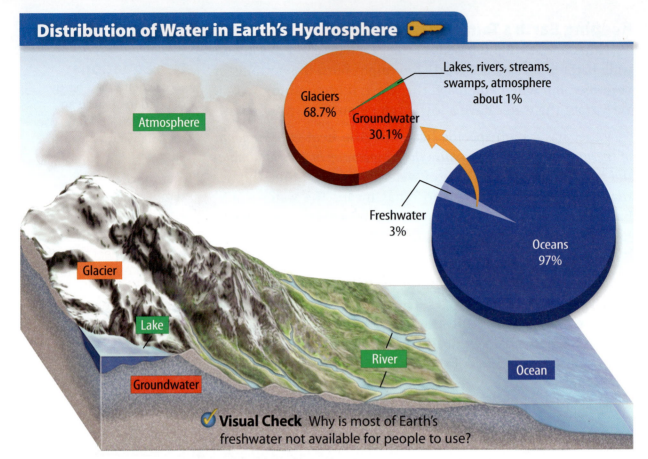

Visual Check Why is most of Earth's freshwater not available for people to use?

Figure 3 About 3 percent of Earth's water is freshwater. Only about 0.001 percent of Earth's water is in the atmosphere.

Water on Earth

You have just read several reasons why water is important for life. You also use water every day for bathing, cooking, and drinking. About 70 percent of Earth's surface is covered by water. How is all this water distributed?

Distribution of Water on Earth

Notice in **Figure 3** that most of Earth's water is in oceans. Only about 3 percent is freshwater (not salty). Freshwater is on Earth's surface, in the ground, or in icecaps and glaciers. Only about 1 percent of all water on Earth is in lakes, rivers, swamps, and the atmosphere.

 Key Concept Check How is water on Earth distributed?

Structure of the Hydrosphere

The **hydrosphere** *is all the water on and below Earth's surface and in the atmosphere.* The many parts of the hydrosphere are shown in **Figure 3**. Water is in oceans, lakes, rivers, and streams and underground. Water beneath Earth's surface is called groundwater. Water vapor, or water in the gaseous state, is in the atmosphere. Clouds are a collection of tiny droplets of water or ice crystals. Ice, or water in the solid state, is in glaciers and ice caps. Earth's frozen water is often called the cryosphere.

WORD ORIGIN

hydrosphere
hydro–
from Greek *hydor*, means "water"
–sphere
from Greek *spharia*, means "ball"

Water Changes State

The only substance that exists in nature in three states—solid, liquid, and gas—within Earth's temperature range is water. It can easily change state within the hydrosphere. For example, in the spring, snow and ice—both solid water—melt to a liquid. When enough thermal energy is added, the liquid water changes to a gas and enters the atmosphere. When water changes from one state to another, thermal energy is either absorbed or released. Thermal energy always moves from an object with a higher temperature to an object with a lower temperature.

Between Solid and Liquid

When thermal energy is added to ice, the water molecules gain energy. If enough thermal energy is added, the ice eventually reaches its melting point and changes to a liquid. The reverse happens if thermal energy is released from liquid water. The molecules begin to lose energy. If the molecules in water lose enough energy, the liquid reaches its freezing point and ice forms.

Between Liquid and Gas

As thermal energy is added to liquid water, the molecules gain energy and eventually reach the boiling point. At the boiling point, water changes to a gas, or water vapor. It takes less energy for molecules at the surface of water to break free from surrounding molecules, as shown in **Figure 4.** Therefore, water at the surface can change to a gas at temperatures below the boiling point and evaporate. **Evaporation** *is the process of a liquid changing to a gas at the surface of the liquid.* When water vapor molecules lose thermal energy, condensation occurs. **Condensation** *is the process of a gas changing to a liquid.*

✓ **Reading Check** Why can evaporation of water occur below water's boiling point?

MiniLab 20 minutes

What happens to temperature during a change of state?

1. Read and complete a lab safety form.
2. Fill a **500-mL beaker** with **crushed ice.**
3. Place the beaker on a **hot plate,** near a **ring stand.** Place a **thermometer** in the beaker about 2.5 cm from the bottom. Use a **clamp** on the ring stand to hold the thermometer in place.
4. In your Science Journal, record the temperature of the ice. Turn the hot plate on medium-high.
5. Record the temperature every minute until 3 minutes after the water starts boiling.

Analyze and Conclude

1. **Identify** When did a change of state occur?
2. **Describe** How did the temperature of the water change as its state changed?
3. **Key Concept** Why is the range of temperatures between the states of water important to life on Earth?

Figure 4 Evaporation occurs only at a liquid's surface.

FOLDABLES

Fold a sheet of paper into a tri-fold book. Record information about the three main phases of the water cycle.

Evaporation
Condensation
Precipitation

Figure 5 Water changes from one state to another as it cycles throughout Earth's hydrosphere.

The Water Cycle

The series of natural processes by which water continually moves throughout the hydrosphere is called the **water cycle**. As water moves through the water cycle, it continually changes state.

Driving the Water Cycle

Two main factors drive the water cycle—the Sun and gravity. Energy from the Sun causes water on Earth's surface to evaporate. The water later falls back to the ground as precipitation. On Earth's surface, gravity moves water from higher to lower areas. Water eventually returns to oceans and other storage areas in the hydrosphere, and the cycle continues.

 Reading Check What two main factors drive the water cycle?

Evaporation

Water on Earth's surface evaporates because energy from the Sun breaks the bonds between water molecules. Liquid water changes into water vapor and enters the atmosphere. As shown in **Figure 5,** evaporation occurs throughout the hydrosphere.

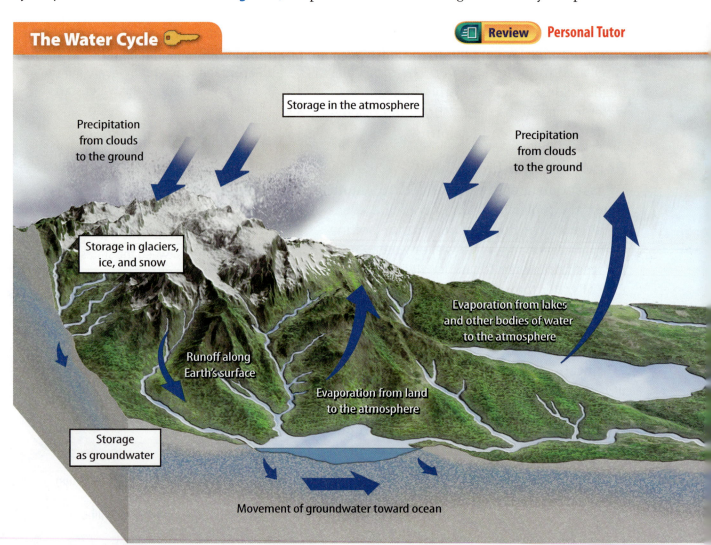

532 • Chapter 15
EXPLAIN

Transpiration *The evaporation of water from plants is called* **transpiration.** Water is absorbed by plants mostly from the ground. When a plant has an abundant water supply or air temperatures increase, plants transpire—they release water vapor into the atmosphere. This usually occurs through the leaves.

Condensation and Precipitation

As water vapor from transpiration and evaporation rises in the atmosphere, it cools and condenses into a liquid. Water vapor condenses around particles of dust in the atmosphere and forms droplets. The droplets combine and form clouds. They eventually fall to the ground as rain. If the temperature is low enough, the water droplets will freeze in the atmosphere and reach Earth's surface as other forms of precipitation such as snow, sleet, or hail.

Runoff and Storage

What happens to the precipitation in **Figure 5** once it reaches Earth's surface? Gravity acts on the precipitation. It causes water on Earth's surface to flow downhill. Water from precipitation that flows over Earth's surface is called runoff. Runoff enters streams and rivers and eventually reaches lakes or oceans. Some precipitation soaks into the ground and becomes groundwater.

Although water is constantly moving through the water cycle, most water remains in certain storage areas for relatively long periods of time. A storage area of the water cycle is called a reservoir. Reservoirs can be lakes, oceans, groundwater, glaciers, and ice caps.

 Key Concept Check Explain the steps as water cycles through Earth's hydrosphere.

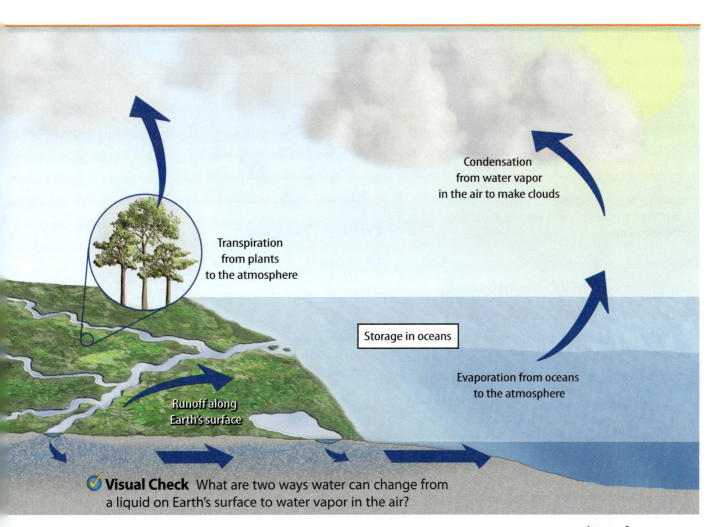

Visual Check What are two ways water can change from a liquid on Earth's surface to water vapor in the air?

Lesson 1 Review

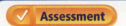 Assessment   Online Quiz

Visual Summary

All organisms on Earth depend on water for survival. Water is a habitat for many organisms.

Water's high specific heat causes large bodies of water to take a long time heating up and cooling down.

The water cycle is a natural process in which water constantly moves throughout the hydrosphere.

FOLDABLES

Use your lesson Foldable to review the lesson. Save your Foldable for the project at the end of the chapter.

What do you think NOW?

You first read the statements below at the beginning of the chapter.

1. A liquid can change to a gas only when the liquid reaches its boiling point.
2. Clouds are made of tiny drops of water.

Did you change your mind about whether you agree or disagree with the statements? Rewrite any false statements to make them true.

Use Vocabulary

1. **Distinguish** between evaporation and transpiration.
2. **Use the term** *hydrosphere* in a complete sentence.
3. **Define** *condensation* in your own words.

Understand Key Concepts

4. **Where is most of the water on Earth?**
   - A. glaciers
   - B. groundwater
   - C. oceans
   - D. rivers
5. **Analyze** What are three reasons water is important to life on Earth?
6. **Name** the two main factors that drive the water cycle.

Interpret Graphics

7. **Identify** the process that occurs at each numbered part of the water cycle below.

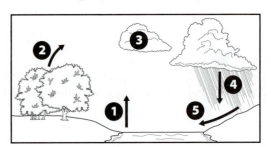

Critical Thinking

8. **Suppose** the amount of water vapor in the atmosphere increased. How would this affect temperatures on Earth's surface? Why?
9. **Evaluate** On a hot day, the water in a swimming pool is much cooler than the cement around the pool. Explain.

Math Skills
Math Practice

10. About how much energy is needed to increase the temperature of 5 kg of sand from 18°C to 32°C if the specific heat of the sand is 190 J/kg • °C?

Oceans on the Rise—Again

CAREERS in SCIENCE

With an eye to the future, a geologist examines past connections between higher sea levels and melting ice sheets.

Way up in the Arctic is the world's largest island—an ice-covered island called Greenland. A vast ice sheet covers much of Greenland. Changes in Earth's climate can have a great effect on this ice sheet. As average global temperatures increase, Greenland's ice sheet is slowly melting along its coastline. If this continues, it could have a big impact on sea levels worldwide.

Greenland's enormous glaciers creep along inch by inch toward the ocean. When they reach the coast, huge chunks of ice break off and crash into the ocean. As the climate warms, the glaciers speed increases, adding more and more ice to polar waters. As more ice enters the ocean, sea levels rise. Scientists estimate that if the Greenland ice sheet melts, sea levels could rise 7 m, or 23 ft. That's enough water to flood coastlines everywhere on Earth, including those with some of the world's largest cities. Imagine New York City under water! Increased flooding also threatens coastal habitats. Animals as well as people would be forced inland. The worst effects would be in a delta—a low-lying area of land where a river flows into a large body of water.

Is this really possible? Scientists, such as Daniel Muhs, know it is possible because it has happened before. Muhs is a geologist with the United States Geological Survey. He investigates rocks for clues about Earth's past. He found a big clue in a limestone wall in the Florida Keys. Today, this wall is several meters above sea level and is filled with fossilized coral. Muhs determined that a coral reef grew there about 125,000 years ago during a warm period when much of the Greenland ice sheet melted. Muhs estimates that sea levels were between 6 m and 8 m higher 125,000 years ago than they are today. This is the same rise in sea levels that other scientists predict would occur if Greenland's ice sheet were to melt again.

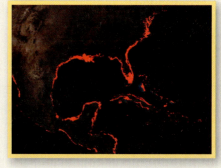

▲ This map shows coastal areas that would be flooded if the sea levels rise 6 m as scientists predict.

Coral lives and grows under water. Muhs shows where the sea level was in the past at this location in Florida. By measuring the height of the coral fossils, he estimates the ocean was once several meters higher than it is today. ▼

▲ A chunk of ice from Greenland's Russell Glacier breaks off and splashes into the ocean. The freshwater in the glacier is no longer available as a source of freshwater when it mixes with seawater. If this continues, the overall amount of freshwater on Earth will decrease.

It's Your Turn

RESEARCH Brainstorm ways that society could respond to rising sea levels. Then, research ways that high sea levels already impact cities and coastlines worldwide. Compare your ideas with real-world solutions and share them with your class.

Lesson 2

Reading Guide

Key Concepts
ESSENTIAL QUESTIONS

- What makes water a unique compound?
- How does water's structure determine its unique properties?
- How does water's density make it important to life on Earth?

Vocabulary

polarity p. 538
cohesion p. 539
adhesion p. 539

 Multilingual eGlossary

The Properties of Water

Inquiry Will they freeze?

A thick layer of ice formed over the water where these penguins swim and hunt. Will the rest of the water freeze? How can plants and animals that live in oceans and lakes survive the winter?

Inquiry Launch Lab

10 minutes

How many drops can fit on a penny?

The structure of a water molecule gives water many unique properties. In this lab, you will explore one property of water—the strong attraction between individual water molecules.

1. Read and complete a lab safety form.
2. Place two **pennies** on a **paper towel.**
3. Use a **dropper** to place 1 drop of **water** at a time on a penny. After 6 drops, closely observe the water on the penny. Try to add more drops, if possible.
4. Use a clean **dropper** to place drops of **rubbing alcohol** one at a time on a different penny. After 6 drops, closely observe the alcohol on the penny. Try to add more drops.

Think About This

1. **Explain** what happened to the water each time you added a drop. What happened when you added the final drop?
2. **Describe** the difference in the shapes of the water and alcohol on the pennies.
3. **Key Concept** Liquids form drops because of the attraction between their particles. Based on this, infer which substance has a stronger attraction between its particles.

Water—A Unique Compound

In Lesson 1 you read that water is the only substance that exists in nature as a solid, a liquid, and a gas. You also read that water has a high specific heat. You might have seen other things that result from water's properties. For example, you have probably noticed that water forms drops if you spill some on a counter, as shown in **Figure 6.** You probably have also seen ice floating in a glass of water or tea. Have you ever dissolved salt in water? Have you ever seen an insect walk on the surface of water?

Water has unusual properties because of its molecules. The properties of water cannot be explained without looking at the way a water molecule is put together. Understanding how water molecules interact with each other and with other materials also helps explain water's unusual properties.

Key Concept Check What makes water a unique compound?

Figure 6 Water forms drops because of strong forces between water molecules.

Lesson 2
EXPLORE
537

Figure 7
The polarity of water molecules is one of the reasons water is so important to life on Earth ▶

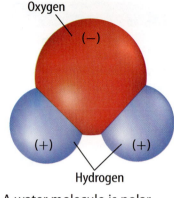

A water molecule is polar because it has a slight charge at each end.

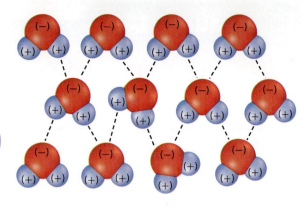

The slightly negative oxygen atom in one water molecule attracts the slightly positive hydrogen atom of another water molecule. This force holds the molecules together.

SCIENCE USE V. COMMON USE

polar
Science Use having opposite ends, which have opposite charges

Common Use relating to Earth's North Pole or South Pole

A Polar Molecule

A water molecule is made up of one oxygen atom and two hydrogen atoms. Look at **Figure 7.** What do you notice about the charges of the atoms? The oxygen atom has a slightly negative charge. The hydrogen atoms have slightly positive charges. The overall charge of a water molecule is neutral. **Polarity** *is a condition in which opposite ends of a molecule have slightly opposite charges, but the overall charge of the molecule is neutral.* Water is a **polar** molecule because the oxygen atom and the hydrogen atoms have slightly opposite charges.

Because of their polarity, water molecules can attract other water molecules. In **Figure 7,** a slightly negative oxygen atom of one water molecule attracts a slightly positive hydrogen atom of another water molecule. Several of water's unique properties are due to its polarity. One of these properties is water's ability to dissolve many different substances.

 Reading Check Describe the polarity of a water molecule.

Water as a Solvent

Water is sometimes called the universal solvent because so many substances can dissolve in it. When table salt, or sodium chloride, is placed in water, it dissolves easily. But how?

Study **Figure 8.** Notice that the positively charged sodium ion (Na^+) of salt is attracted to the negatively charged oxygen atom of the water molecule. The negatively charged chloride ion (Cl^-) of salt is attracted to the positively charged hydrogen atom of the water molecule. These attractions cause the sodium and chloride ions to break apart in water, or dissolve. Many substances that are important to life processes are dissolved in water within cells, blood, and plant tissues.

▲ **Figure 8** Ionic compounds, such as table salt, (NaCl) can easily dissolve in water because water is polar.

Cohesion and Adhesion

How can the water strider shown in **Figure 9** walk across the surface of water? You've read that water molecules attract each other because of their polarity. This attraction is called cohesion. **Cohesion** *is the attraction among molecules that are alike.* Some insects can walk on the surface of water because the attractions among water molecules is stronger than the attraction of gravity on the insect. The ability to put more drops of water than alcohol on the penny in the Launch Lab also demonstrates cohesion.

Adhesion *is the attraction among molecules that are not alike.* You might be familiar with one example of adhesion—the formation of a curved surface, called a meniscus, on a liquid in a test tube, as in **Figure 9.** Notice that the water molecules in contact with the sides of the test tube stick to the glass, causing the curved surface across the top of the water.

Water moves from the roots of a plant to its leaves as a result of both cohesion and adhesion. As a water molecule evaporates from the surface of a leaf, it pulls another water molecule up into its place. Water molecules stick to the cells within the plant. This keeps gravity from pulling water back down toward the roots.

 Key Concept Check Name some ways that water's structure determines its unique properties.

WORD ORIGIN
cohesion
from Latin *cohaerere*, means "to stick together"

FOLDABLES
Fold a sheet of paper into a two-tab concept map. Label it as shown. Use your book to summarize information about water and its properties.

Properties of Water
| Cohesion | Adhesion |

Cohesion and Adhesion

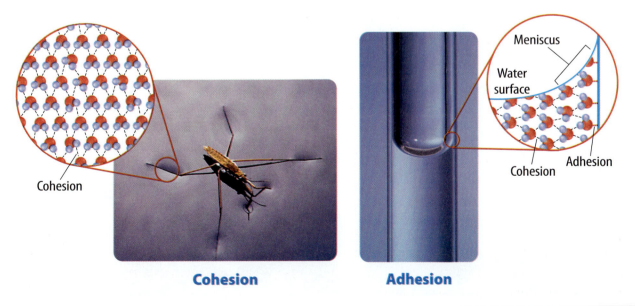

Figure 9 Cohesion is responsible for molecules of water sticking together. Adhesion is responsible for water molecules sticking to other surfaces.

Cohesion

Adhesion

REVIEW VOCABULARY
density
mass per unit volume of a material

ACADEMIC VOCABULARY
establish
(verb) to make; to put in place

Figure 10 Water is denser than ice because molecules are packed more closely in water than in ice.

Density

Have you ever wondered why ice cubes float in a glass of water? Water that freezes on a lake also floats. Even huge icebergs like the one in **Figure 10** float in the ocean. Ice floats in liquid water because of an important property—**density**.

The density of a material increases when the particles in the material get closer together. When most liquids freeze, their particles get closer together. The solid that forms is denser than the liquid. For example, recall that lava is molten, or liquid, rock. As lava cools, the particles get closer together. Therefore, the rock that forms from the lava is denser than lava. The rock will sink if placed in the lava.

Water's Unusual Density

If liquids tend to get denser as they freeze, why does ice float in water? Just like any other liquid, as water cools, the molecules lose energy and pack tightly together. However, when water cools to 4°C the molecules begin to move farther apart. Forces among the molecules cause the molecules to spread out and **establish** themselves in a six-sided pattern. When the water molecules freeze, there is space between them. A cube of ice has fewer water molecules than the same volume of water. Therefore, ice is less dense than water, as shown in **Figure 10**.

Reading Check Why is it unusual that ice floats?

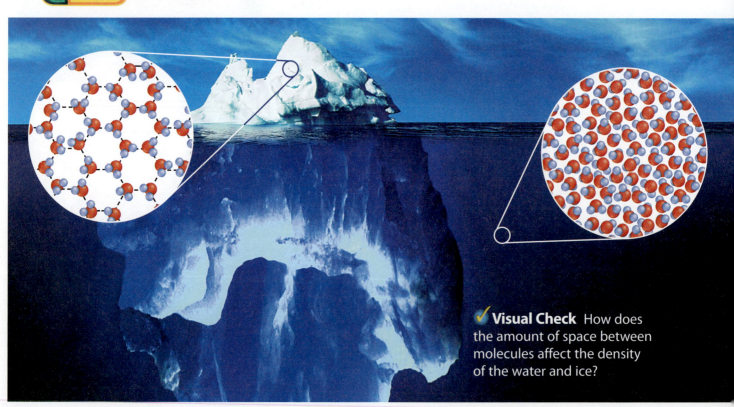

Visual Check How does the amount of space between molecules affect the density of the water and ice?

Figure 11 The density of liquid water is greater than that of ice. The density of liquid freshwater is greatest at a temperature of 4°C.

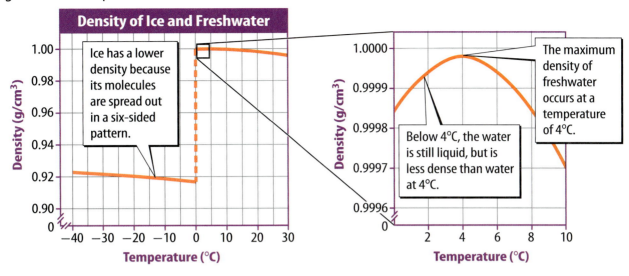

Density and Temperature

To understand more about the unusual density of water, study the two graphs in **Figure 11**. Both graphs illustrate how density changes as the temperature changes. The graph on the left shows the density of both water and ice. The graph on the right is a close-up view of water's density change.

The Density of Ice You can compare the density of ice and water in the graph on the left. The density of ice is much lower than the density of water. Recall that the molecules in ice are more spread out than in water. This explains why ice floats.

The Density of Water Only the density of water is represented in the graph on the right. This graph shows that water is most dense at a temperature of 4°C. Remember that the freezing point of water is 0°C. This means that water between 0°C and 4°C is liquid, but it is less dense than water at 4°C. As you will read on the next page, density of water is important for the survival of life in the water.

 Reading Check How does the density of water at 0°C differ from the density of liquid water at 4°C?

Inquiry MiniLab 20 minutes

Is every substance less dense in its solid state?

Is olive oil less dense in its solid state?

1. Read and complete a lab safety form.
2. Pour 20 mL of **liquid olive oil** into a 50-mL **graduated cylinder**.
3. Form a hypothesis about whether olive oil is more dense as a solid or as a liquid.
4. Drop a chunk of **solid olive oil** into the liquid olive oil. Record what happens in your Science Journal.

Analyze and Conclude

1. **Analyze** Is liquid olive oil or solid olive oil more dense? How do you know this?
2. **Key Concept** How do the densities of solid and liquid olive oil differ from those of solid and liquid water?

Figure 12 🔑 Fish and other organisms in a lake can survive in winter because the water remains below a layer of ice.

① When surface water cools to 4°C and sinks, warmer, less dense water rises. This process continues until all water is 4°C and equally dense.
Air: −5°C

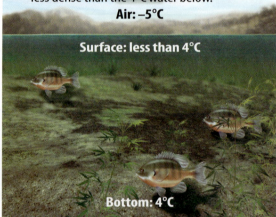
Surface: 4°C
Bottom: greater than 4°C

② The air cools the surface water below 4°C. The cooler water remains at the surface because it is less dense than the 4°C water below.
Air: −5°C

Surface: less than 4°C
Bottom: 4°C

③ Ice forms at 0°C and remains at the surface because it is less dense than liquid water. The ice insulates the water below.
Air: −5°C
Ice: 0°C
Surface: 4°C
Bottom: 4°C

Visual Check What is the temperature of the water below the ice?

The Importance of Water's Density

You have just read about two important features of water's density:

- The density of ice is lower than the density of water.
- The density of freshwater is greatest at 4°C.

Imagine a lake in winter, as shown in **Figure 12**. How is the density of ice and water important to the survival of some organisms on Earth?

① In winter, cold air above a lake cools the surface water. When the surface water cools to 4°C, it reaches its maximum density and is more dense than the water below it. As a result, the surface water sinks while pushing the warmer water to the surface. Once the warmer water reaches the surface, the air cools this water to 4°C. Again, the water becomes more dense and sinks.

② Eventually, all the water in the lake cools to 4°C and is equally dense. However, the air above the water continues to cool the surface water. The temperature of the surface water drops below 4°C, and the density begins to decrease. The colder surface water is less dense than the 4°C water below it and stays on top. All the water below the surface water remains at 4°C and maximum density.

③ When the surface water of the lake cools to 0°C, it changes to ice. The density of the ice decreases further, and it continues to float. Ice on the surface insulates the water below it. Aquatic organisms can survive cold, winter months because beneath the ice, water remains a liquid at 4°C. If water froze from the bottom of a lake to the top, organisms living in the lake would freeze along with the water.

 Key Concept Check How does water's density make it important to life on Earth?

Lesson 2 Review

 Assessment   Online Quiz

Visual Summary

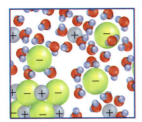
Water can dissolve many substances because a water molecule is polar.

Cohesion is an important property of water molecules. Molecules at the surface of water have enough cohesion that some insects can walk on the surface of water.

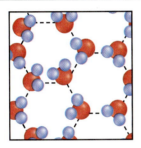
Ice is less dense than water because as water freezes the molecules spread out in a six-sided pattern.

FOLDABLES

Use your lesson Foldable to review the lesson. Save your Foldable for the project at the end of the chapter.

What do you think NOW?

You first read the statements below at the beginning of the chapter.

3. Water molecules can attract other water molecules.

4. Ice has a greater density than water.

Did you change your mind about whether you agree or disagree with the statements? Rewrite any false statements to make them true.

Use Vocabulary

1 A property in which opposite ends of a molecule are slightly charged is _____.

2 **Distinguish** between adhesion and cohesion.

Understand Key Concepts

3 Which has the highest density?
  A. water at 0°C   C. water at 6°C
  B. water at 4°C   D. water at 8°C

4 **Relate** the structure of water molecules to water's unique properties.

5 **Describe** how water's unusual density is important to organisms in a lake in winter.

Interpret Graphics

6 **Organize Information** Copy and fill in the graphic organizer below to describe examples of adhesion and cohesion.

Adhesion	
Cohesion	

7 **Analyze** Use the graph below to describe how the density of water changes if the water temperature is increased—between 0°C and 4°C; at 4°C; between 4°C and 10°C.

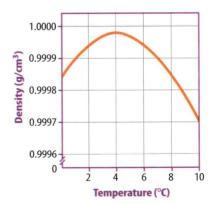

Critical Thinking

8 **Compose** Chris placed two cubes in water. Compose a statement that describes why one cube sank and the other floated. Use the term *density* in your answer.

Lesson 2 • EVALUATE • 543

Inquiry Skill Practice: Cause and Effect

15 minutes

Why is liquid water denser than ice?

Ice floating on a lake in winter is important to the survival of organisms in the lake. Ice floats because its density is lower than the density of water. What is the cause of this difference in density? What effect does the structure of water have on its density?

Materials

modeling clay (two colors)

24 toothpicks

Safety

Learn It

An important part of science is being able to understand **cause and effect** relationships based on observations. Cause and effect is the concept that an event will produce a certain response. You will use models to observe the cause-and-effect relationship between the structure of water and its density.

Try It

1. Read and complete a lab safety form.

2. Use toothpicks and modeling clay to model 12 water molecules. One color of clay represents oxygen atoms. The other color represents hydrogen atoms.

3. The molecules that make up water move freely and are disorganized. Use six of your models and show the water molecules closely arranged.

4. Recall that water molecules are polar. The oxygen atom of one molecule is attracted to a hydrogen atom of another molecule. However, atoms that are alike push away from each other. As water freezes, these forces cause water molecules to form a six-sided pattern. Use the remaining six models to form one of these six-sided patterns, as shown below.

Apply It

5. **Identify** Which has more empty space between molecules, the model of liquid water or the model of ice? What causes this empty space?

6. **Key Concept** Based on your observations, what effect does the structure of water molecules have on the density of liquid water and ice?

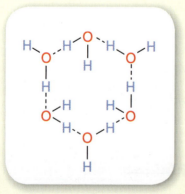

Lesson 3

Water Quality

Reading Guide

Key Concepts 🔑
ESSENTIAL QUESTIONS
- Why is water quality important?
- How is water quality tested and monitored?

Vocabulary
water quality p. 546
point-source pollution p. 547
nonpoint-source pollution p. 547
nitrate p. 549
turbidity p. 549
bioindicator p. 550
remote sensing p. 550

 Multilingual eGlossary

Inquiry Clean Water?

The water in the pond on this glacier looks clean enough to drink, but is it? Can you always tell just by looking at water whether it is clean? How do human activities affect the quality of water?

Inquiry Launch Lab

10 minutes

How can you test the cloudiness of water?

All lakes and ponds contain sediment, but too much sediment is one way that water can become cloudy. Cloudy water can sometimes be a problem for organisms that live in the lakes and ponds.

1. Read and complete a lab safety form.
2. Tie a **bolt** onto the end of a **string.** Lower the bolt into a **bucket** of **water.** Record notes in your Science Journal.
3. Add **soil** to the water until the water is cloudy. Use a **long-handled wooden spoon** to stir the sediment.
4. Lower the bolt to the same depth as step 1. Record your observations.

Think About This

1. How did your observation of the bolt change after you added soil to the water?
2. **Key Concept** How might scientists use a similar method to study the cloudiness at different depths of a lake or pond?

Human Effect on Water Quality

Suppose you go to the beach, looking forward to swimming and playing in the waves. When you arrive, you find a warning sign like the one in **Figure 13.** What might the sign tell you about the quality of the water and the health of the organisms that live in it?

Water quality is the chemical, biological, and physical status of a body of water. It also describes the water's characteristics, such as the amount of oxygen and nutrients in the water, the type and number of organisms living in the water, and the amount of sediment in the water. All of these characteristics are important to the health of aquatic organisms.

Many natural processes, such as seasonal temperature changes and the weathering of rock and soil, affect the quality of water. Human activities can also affect water quality. Pollution from factories and automobiles eventually reaches rivers, lakes, wetlands, and oceans. Deforestation, which removes large numbers of trees, can lead to increased soil erosion. In addition, when it rains, runoff carries soil and other materials into streams and rivers, changing the quality of the water.

Reading Check What are some ways human activities affect water quality?

Figure 13 This sign warns about the quality of Earth's water.

◀ **Figure 14** The water pollution shown here is point-source pollution because its source is known.

Visual Check Can you identify the origin of pollution in this photo?

Point-Source Pollution

How are sources of pollution classified? The wastewater flowing out of the pipe in **Figure 14** is an example of point-source pollution. **Point-source pollution** is *pollution that can be traced to one location,* such as a drainpipe or a smokestack.

The pollution in **Figure 14** is coming from a factory—a common origin of point-source pollution. Another origin of point-source pollution is sewage treatment plants. In many older sewer systems, water from precipitation is mixed with wastewater before being treated. During heavy rainstorms, the sewage treatment plant cannot process the excess water. As a result, storm water, along with untreated sewage, is released directly into nearby bodies of water.

Nonpoint-Source Pollution

Pollution that cannot be traced to one location is **nonpoint-source pollution.** Runoff from large areas, such as lawns, roads, and urban areas, is considered nonpoint-source pollution. As shown in **Figure 15,** the runoff might flow into rivers or streams. It eventually reaches areas of water storage, such as a wetland, groundwater, or the ocean. The runoff might contain natural and human-made pollutants such as sediment, fertilizers, and oil.

Like point-source pollution, nonpoint-source pollution can lower water quality. It can lead to changes in water, which can harm aquatic organisms. Certain types of fish might not be safe for humans to eat because they have high levels of toxins in their tissues. Nonpoint-source pollution might also affect drinking water.

Key Concept Check Why is water quality important?

FOLDABLES

Make a half-book from a sheet of paper. Use your book to organize your notes about the effect different types of pollution have on water quality.

Figure 15 Much water pollution is from nonpoint-source pollution. ▼

Lesson 3
EXPLAIN

Dissolved Oxygen

Figure 16 The aerator in this fish tank releases bubbles that help keep the water moving throughout the tank. This allows oxygen in the air to continually dissolve in the water at its surface.

Testing Water Quality

Scientists examine water quality using a variety of tests. These tests include measuring levels of dissolved gases, temperature, acidity, and cloudiness. Studying the numbers or the health of certain aquatic organisms is another way scientists measure water quality. Using photos taken from the air or space can also help scientists compare the quality of water over time.

Dissolved Oxygen

Why can fish breathe under water but people cannot? Like the air you breathe, water in oceans and lakes contains oxygen. Some of this oxygen is dissolved in the water. Fish, such as the ones in **Figure 16,** use gills to take in this oxygen they need to survive.

The level of dissolved oxygen affects water quality. If the oxygen level in a lake or stream becomes too low, fish might not be able to survive. Different factors can affect oxygen levels. For example, the release of certain chemicals in water can cause an overgrowth of algae. When the algae die, the decay process uses a large amount of oxygen. The oxygen level in the water can drop so low that fish die.

Water Temperature

Many aquatic organisms are also sensitive to changes in water temperature. Coral bleaching is the whitening of coral due to stress in the environment, such as an increase in water temperature or increased exposure to ultraviolet radiation. It is an event that leads to the death of large areas of coral reefs and is often triggered by a temperature increase in water as little as 2°C. As water temperature increases, the amount of oxygen that can dissolve in water decreases. This means that as water temperature increases, there is less oxygen in the water, which can be harmful to aquatic animals.

Reading Check How would cooling water affect the level of dissolved oxygen?

Inquiry MiniLab   20 minutes

How do oxygen levels affect marine life?

The table below contains every other month's average dissolved oxygen levels in the Chesapeake Bay from 1985 through 2002.

On **graph paper,** make a line graph using the data in the table.

Month	Dissolved Oxygen
January	10.0 mg/L
March	10.0 mg/L
May	5.0 mg/L
July	1.5 mg/L
September	3.0 mg/L
November	7.0 mg/L

Analyze and Conclude

1. **Describe** the pattern of dissolved oxygen levels throughout the entire year.
2. **Key Concept** Blue crabs need at least 3 mg/L of dissolved oxygen to survive. Infer during which month(s) the levels of dissolved oxygen might affect the population of blue crabs in the Bay.

Nitrates

Compounds that contain the nitrate ion can be harmful to the environment. A **nitrate** *is a nitrogen-based compound often used in fertilizers.* Runoff from fertilizers used in landscaping and farming contribute to high concentrations of nitrate found in water. This can cause an algal bloom, in which the algae population increases at a rapid rate, as shown in **Figure 17.** Algae growing on the water's surface can block light needed by plants growing at greater depths, causing them to die. The algae can die too. When the algae die, oxygen levels in the water can decrease, producing a very unhealthy ecosystem.

 Reading Check What is an algal bloom?

Acidity

When scientists work in a lab with substances that are strong acids or strong bases, they have to be extremely careful. These substances can be harmful. Strong acids and bases can also be harmful to animals and plants that live in water. Long-term changes in the acidity of water can affect the entire ecosystem. Some fish might not be able to survive. Even if some organisms survive in acidic water, their food sources might not.

Turbidity

A measure of the cloudiness of water, from sediments, microscopic organisms, or pollutants, is **turbidity** (tur BIH duh tee). As the amount of matter floating in water increases, the turbidity increases. Also, the distance light can penetrate into water decreases. Turbidity affects organisms that need light to undergo photosynthesis. High turbidity can also affect filter-feeding organisms. The structures these organisms use to filter food from water can get clogged with sediment. The organisms could die from lack of food. Turbidity is measured using a device called a Secchi disk, shown in **Figure 18.**

▲ **Figure 17** Nitrates from farm fertilizer flow into this stream, causing an algal bloom.

WORD ORIGIN
turbidity
from Latin *turbidus*, means "disturbance"

Figure 18 A Secchi disk is used to measure turbidity. The farther down in the water the disk is visible, the lower the turbidity of the water. ▼

Measuring Turbidity

Bioindicators

An organism that is sensitive to environmental conditions and is one of the first to respond to changes a **bioindicator.** Bioindicators alert scientists to changes in the level of oxygen, nutrients, or pollutants in the water. For example, the presence of stoneflies, small insects that live on the bottom of streams, usually indicates good water quality. Stoneflies cannot survive when oxygen levels in water are too low.

Larger organisms, such as fish, also can be used as bioindicators. The number of fish species in different locations in Florida are shown in the graph in **Figure 19.** The different species are classified as tolerant, moderately tolerant, and not tolerant of pollution. When species that are not tolerant of pollution are missing from water, this can indicate poor water quality.

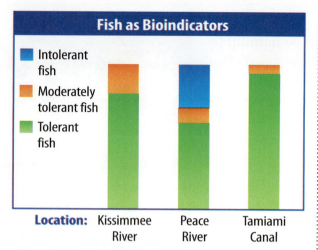

▲ **Figure 19** The presence of intolerant fish indicates that Peace River has good water quality.

✓ **Visual Check** Which area of water most likely has the worst water quality?

Remote Sensing

The collection of data from a distance is called **remote sensing.** Remote sensing data can be collected through photos taken from the air or images taken from satellites. Scientists use remote sensing data to monitor changes in water storage on Earth, such as melting glaciers. Images from satellites can be used to compare water in the same area over time.

Data collected through remote sensing can be used to make inferences about water quality. Notice in **Figure 20** that the water in Pamlico Sound was loaded with sediment after Hurricane Floyd in 1999. Heavy rain from the hurricane led to large amounts of runoff from land. The nutrients in the runoff caused an overgrowth of algae. When the algae died, oxygen levels decreased. Population levels of blue crabs, oysters, and clams all decreased as a result.

Figure 20 The top photograph shows Pamlico Sound before Hurricane Floyd. The bottom photograph shows the sediment deposited by runoff into Pamlico Sound after the hurricane's heavy rainfall. ▼

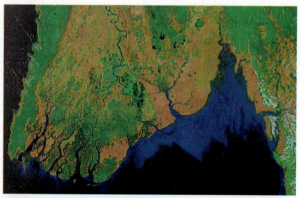

✓ **Key Concept Check** Name several ways water quality is tested and monitored.

Lesson 3 Review

Assessment — Online Quiz
Inquiry — Virtual Lab

Visual Summary

Water quality is the chemical, biological, and physical status of a body of water. Sources of pollution are not always obvious.

Various factors can cause a decreased level of dissolved oxygen in water. This can harm aquatic organisms.

High turbidity is another factor that can harm aquatic organisms.

FOLDABLES

Use your lesson Foldable to review the lesson. Save your Foldable for the project at the end of the chapter.

What do you think NOW?

You first read the statements below at the beginning of the chapter.

5. Factories are responsible for almost all water pollution.
6. Changes in the types of organisms living in water can be a sign of changes in the quality of the water.

Did you change your mind about whether you agree or disagree with the statements? Rewrite any false statements to make them true.

Use Vocabulary

1. A measure of the cloudiness of water is called _____.
2. **Use the term** *bioindicator* in a complete sentence.
3. **Define** the terms *point-source pollution* and *nonpoint-source pollution* in your own words.

Understand Key Concepts

4. What is a way that water changes as the temperature of the water increases?
   A. Acidity decreases.
   B. Acidity increases.
   C. Oxygen level decreases.
   D. Oxygen level increases.
5. **Explain** how a change in water acidity can affect organisms living in a lake.
6. **Decide** A scientist is monitoring the water quality of two lakes. One lake contains a high level of intolerant fish. The other lake contains a low level of intolerant fish. Which lake most likely has better water quality? Why?

Interpret Graphics

7. **Sequence** Draw a graphic organizer like the one below to sequence how an overgrowth of algae in a lake can kill fish.

   | Chemicals are released into water. | → | → | → |

Critical Thinking

8. **Predict** A river recently experienced an algal bloom. There are no stoneflies in the river. What might a scientist find about the level of nitrates, the level of oxygen, or the turbidity of the water?
9. **Recommend** A scientist in New York wants to study changes in the size of glaciers in Antarctica over the next ten years. What type of remote sensing could she use?

Lesson 3
EVALUATE
551

Inquiry Lab

45 minutes

Temperature and Water's Density

Materials

food coloring

hot plate

250-mL beakers (3)

ice

stirring rods (2)

droppers (2)

heat-resistant glove

Safety

If water were like most substances, ice would sink in liquid water, and underwater organisms would die as lakes and ponds froze completely in winter. But the properties of water are different from most substances, and these properties are important for life on Earth. In this lab, you will investigate the relationship between water's temperature and its density. If one material floats in another, the material that floats has a lower density.

Question

What effect does temperature have on water's density?

Procedure

1. Read and complete a lab safety form.
2. Copy the data table in your Science Journal.
3. Stir 3 drops of blue food coloring into 150 mL of water in a beaker. Stir in ice until the water is cold.
4. Based on your observations, conclude which has a lower density—the ice or the cold water. Record your observations and conclusion in your data table.
5. Stir 3 drops of red food coloring into 150 mL of water in a beaker. Heat the water on a hot plate until the water is warm but not boiling.
6. Place a small amount of ice in the warm water. Observe whether the ice floats. Conclude whether the ice or warm water has a lower density. Record your observations and conclusion.

What was compared?	Observations	Conclusions
Ice and cold water		
Ice and warm water		
Cold water		
Room temperature water		
Warm water		

Chapter 15
EXTEND

Form a Hypothesis

7 Think about what you have observed about the relationship of temperature to the density of water. Form a hypothesis about the differences in density of cold water, room temperature water, and warm water.

Test Your Hypothesis

8 Design an investigation to test your hypothesis about the relationship of density to the temperature of water.
⚠️ *Use a heat-resistant glove to handle the heated glass beaker and stirring rod.*

9 Place 150 mL of room-temperature water into a beaker.

10 Carefully add several drops of warm, dyed-red water into the room temperature water. Then add several drops of cold, dyed-blue water into the room-temperature water. Record observations in your data table.

Lab Tips

☑ Use a dropper to place small amounts of one temperature of water into water which has a different temperature.

☑ Be sure the water in a beaker is as still as possible before placing a different temperature of water in it.

Analyze and Conclude

11 List the following from least dense to most dense: ice, cold water, room-temperature water, warm water.

12 **Conclude** Write a statement that describes how differences in temperature cause differences in the density of water.

13 **The Big Idea** Explain the effect of temperature and density of water on underwater organisms.

Communicate Your Results

Create a poster that explains how the temperature of water is related to its density. Include colorful drawings to illustrate your observations.

 **Extension**

The density of water depends on its temperature. However, in ocean water, differences in the saltiness of water can cause differences in density. Design an experiment that tests this effect.

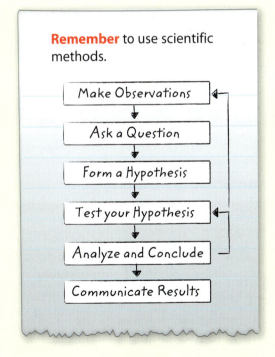

Remember to use scientific methods.

Make Observations → Ask a Question → Form a Hypothesis → Test your Hypothesis → Analyze and Conclude → Communicate Results

Chapter 15 Study Guide

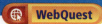

 Water cycles throughout Earth's hydrosphere and is necessary for the survival of all living things.

Key Concepts Summary

Lesson 1: The Water Planet

- All organisms on Earth depend on water. Water regulates Earth's temperature.
- Water provides a stable temperature for aquatic organisms because of its high **specific heat**.
- Water is in the **hydrosphere**—on and below Earth's surface and in the atmosphere.
- Water moves through the **water cycle** by **evaporation, transpiration, condensation,** precipitation, and runoff.

Vocabulary

specific heat p. 529
hydrosphere p. 530
evaporation p. 531
condensation p. 531
water cycle p. 532
transpiration p. 533

Lesson 2: The Properties of Water

- Water is the only substance that exists naturally as a solid, a liquid, and a gas on Earth.
- Because of its **polarity,** water dissolves many substances.
- Together, **cohesion** and **adhesion** allow water to transport nutrients and wastes within plants.
- Since the **density** of ice is less than that of water, ice floats and insulates the water below. This allows aquatic organisms to survive in the winter.

polarity p. 538
cohesion p. 539
adhesion p. 539

Lesson 3: Water Quality

- **Water quality** affects the health of humans and aquatic organisms. The quality of water can be harmed by **point-source pollution** or by **nonpoint-source pollution.**
- The quality of water can be tested by monitoring levels of dissolved oxygen, temperature, **nitrates,** acidity, **turbidity,** and **bioindicators. Remote sensing** is one method of monitoring.

water quality p. 546
point-source pollution p. 547
nonpoint-source pollution p. 547
nitrate p. 549
turbidity p. 549
bioindicator p. 550
remote sensing p. 550

Study Guide

Review
- Personal Tutor
- Vocabulary eGames
- Vocabulary eFlashcards

Chapter Project

Assemble your lesson Foldables as shown to make a Chapter Project. Use the project to review what you have learned in this chapter.

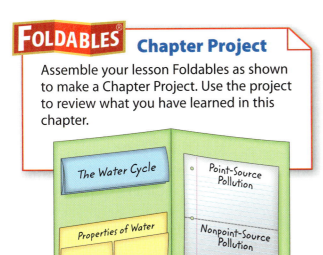

Use Vocabulary

1. Water moves through Earth's _____ by a process called the water cycle.

2. The process of water changing to a gas at its surface is _____.

3. Slightly opposite charges on opposite ends of water molecules cause the _____ of water.

4. The attraction between molecules that are alike is called _____.

5. The chemical, biological, and physical status of a body of water is _____.

6. An organism that is sensitive to environmental conditions and is one of the first to respond to changes is a(n) _____.

Link Vocabulary and Key Concepts

Concepts in Motion — Interactive Concept Map

Copy this concept map, and then use vocabulary terms from the previous page to complete the concept map.

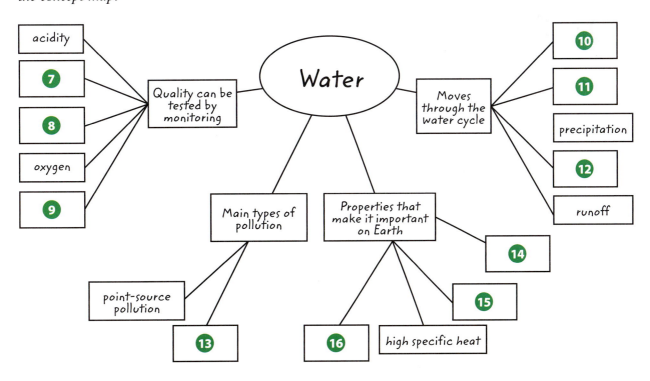

Chapter 15 Study Guide • 555

Chapter 15 Review

Understand Key Concepts

1. The atmosphere has the highest concentration of which greenhouse gas?
   A. carbon dioxide
   B. carbon monoxide
   C. methane
   D. water vapor

2. Which main factors drive the water cycle?
   A. gravity and precipitation
   B. gravity and the Sun's energy
   C. precipitation and evaporation
   D. the Sun's energy and evaporation

3. The diagram below shows the distribution of freshwater on Earth.

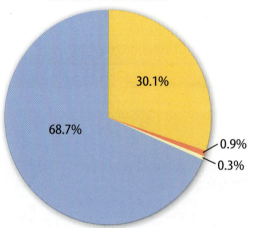

   Earth's Freshwater

   According to the graph and what you have read in this chapter, about how much of Earth's freshwater is in places other than glaciers, icebergs, and groundwater?
   A. 0.3%
   B. 1.2%
   C. 68.7%
   D. 98.8%

4. What is the freezing point of water?
   A. −2°C
   B. 0°C
   C. 4°C
   D. 10°C

5. Which BEST describes the diagram below?

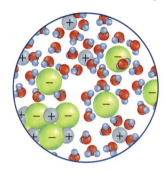

   A. Sodium and chloride ions are adhering to each other.
   B. Sodium and chloride ions are sinking in water.
   C. Sodium chloride is dissolving in water.
   D. Sodium chloride is floating in water.

6. Which property of water is most responsible for an insect being able to walk on the surface of a pond?
   A. adhesion
   B. cohesion
   C. density
   D. transpiration

7. What causes an algal bloom?
   A. a very high acidity level
   B. a very low turbidity level
   C. too much nitrate in the water
   D. too much oxygen in the water

8. Which is nonpoint-source pollution?
   A. leakage from a sewage treatment plant
   B. an oil spill from a tanker ship
   C. runoff from an urban area
   D. warm water from a factory drainpipe

9. Which can be used to measure the level of turbidity of water?
   A. Erlenmeyer flask
   B. microscope
   C. Secchi disk
   D. remote sensing

556 • Chapter 15 Review

Chapter Review

Assessment — Online Test Practice

Critical Thinking

10 Explain how the high specific heat of water is important to living things on Earth.

11 Imagine How would life on Earth change if water did not naturally exist in all three states in the range of temperatures on Earth?

12 Design a demonstration that compares an effect of water's high specific heat to other substances, such as soil or asphalt.

13 Cause and Effect Copy and fill in the graphic organizer below to list a cause and several effects of water's ability to dissolve many substances.

14 Evaluate Detergent breaks the bonds between water molecules. This helps remove grease and oil stains from clothes in the washing machine. However, detergent can enter rivers and lakes in wastewater and runoff. How can this affect the organisms that live in these habitats?

15 Construct a flow chart that explains how the deforestation of an area can affect the water quality of a nearby river.

16 Illustrate why water is a polar molecule.

17 Give an example of how scientists use bioindicators to monitor water quality.

Writing in Science

18 Design a four-page brochure in which you describe and illustrate different ways that human activities affect water quality. Be sure to include ways that human activities both benefit and harm water quality.

REVIEW THE BIG IDEA

19 What role does water play in regulating Earth's temperature?

20 The photo below shows animals that live on the dry grasslands of Africa. Why is water so important to the animals?

Math Skills

Review Math Practice

Use Equations

Substance	Specific Heat (J/kg · °C)
Water (H_2O)	4186
Hard plastic	400
Copper (Cu)	90

21 One kilogram of water, plastic, and copper at room temperature receive the same amount of energy from the Sun over a 10 min period. Which material will have the smallest increase in temperature? Explain.

22 How much energy is needed to warm 8.0 kg of copper from 120°C to 145°C?

23 Two kilograms of a substance needs 20,000 J of energy to warm from 200°C to 300°C. What is the specific heat of the substance? Use this form of the equation:

$$\text{Specific heat} = \frac{\text{energy}}{(\text{mass} \times \text{temperature change})}$$

Chapter 15 Review • 557

Standardized Test Practice

Record your answers on the answer sheet provided by your teacher or on a sheet of paper.

Multiple Choice

1. Which is point-source pollution?
   A acid rain
   B broken drainpipe
   C field runoff
   D weathering rock

Use the diagram below to answer questions 2 and 3.

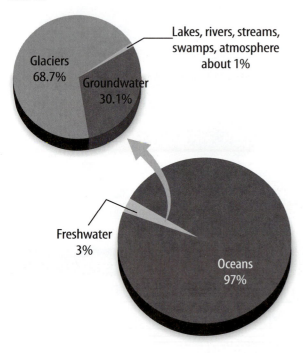

2. According to the graphs, approximately how much of Earth's water resides in glaciers?
   A 2 percent
   B 3 percent
   C 30 percent
   D 68 percent

3. What is the ratio of freshwater to saltwater on Earth?
   A 3:97
   B 3:100
   C 97:3
   D 97:100

4. What property of the molecules in ice makes ice float on water?
   A They are farther apart than water molecules.
   B They are much larger than water molecules.
   C They contain more oxygen atoms than water molecules.
   D They move more quickly than water molecules.

Use the diagram below to answer question 5.

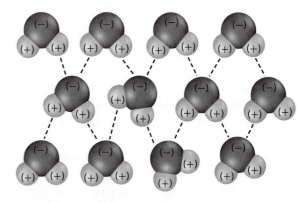

5. What property of water molecules does the diagram illustrate?
   A consistency
   B layering
   C neutrality
   D polarity

6. Which is the physical, chemical, and biological status of a body of water?
   A its density
   B its quality
   C its specific heat
   D its volume

558 • Chapter 15 Standardized Test Practice

Standardized Test Practice

Assessment
Online Standardized Test Practice

Use the diagram below to answer question 7.

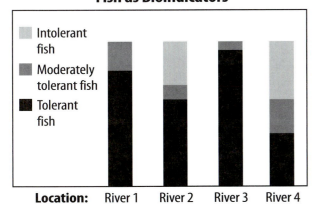

Fish as Bioindicators

- Intolerant fish
- Moderately tolerant fish
- Tolerant fish

Location: River 1, River 2, River 3, River 4

7. In the graph above, which has the best water quality?
   - A river 1
   - B river 2
   - C river 3
   - D river 4

8. When a lake freezes in winter, what happens beneath the ice layer?
   - A Organisms freeze at 4°C.
   - B The water at the bottom turns to ice.
   - C Warm water sinks to the bottom.
   - D Water remains liquid at 4°C.

9. Which explains why water in a cylinder forms a meniscus across the top?
   - A adhesion
   - B density
   - C specific heat
   - D turbidity

Constructed Response

Use the table below to answer question 10.

Stage	Description
Condensation	
Evaporation	
Precipitation	
Runoff	
Storage	

10. In the table above, describe each stage of the water cycle and where it occurs.

Use the table below to answer questions 11 and 12.

Factor	Effect
Acidity	
Dissolved oxygen	
Nitrates	
Temperature	
Turbidity	

11. Explain the effect each factor in the table above has on water quality.

12. How does human activity contribute to the effects these factors have on water quality? Give two examples.

NEED EXTRA HELP?												
If You Missed Question...	1	2	3	4	5	6	7	8	9	10	11	12
Go to Lesson...	3	1	1	2	2	3	3	2	2	1	3	3

Chapter 16

Oceans

 What are characteristics of oceans, and why are oceans important?

Inquiry What makes waves so powerful?

Have you ever felt the power of an ocean wave? Oceans are large and powerful, and they can be dangerous. They are also important. Oceans contain valuable resources, and they affect Earth's climate and weather.

- What causes ocean waves and currents? How do oceans affect weather and climate?
- How are oceans threatened?
- What are the characteristics of oceans, and why are oceans important?

Get Ready to Read

What do you think?
Before you read, decide if you agree or disagree with each of these statements. As you read this chapter, see if you change your mind about any of the statements.

1. Oceans formed about 4 billion years ago.
2. The seafloor is flat.
3. Waves move water particles from one location to another.
4. The wind causes tides.
5. Ocean currents occur on the surface and below the surface.
6. Ocean currents affect climate and weather.
7. Most pollution in the oceans originates on land.
8. Global climate change has no effect on marine organisms.

Lesson 1

Reading Guide

Key Concepts 🔑
ESSENTIAL QUESTIONS

- Why are the oceans salty?
- What does the seafloor look like?
- How do temperature, salinity, and density affect ocean structure?

Vocabulary
salinity p. 565
seawater p. 565
brackish p. 565
abyssal plain p. 566

🅖 Multilingual eGlossary

🎬 Video BrainPOP®

Composition and Structure of Earth's Oceans

Inquiry What's down there?

Conditions change with depth in the ocean. Scientists study different layers of the ocean by diving in submersibles—tiny submarines capable of withstanding extreme pressure at great depths. How do you think the ocean changes with depth?

562 Chapter 16
ENGAGE

Inquiry Launch Lab

15 minutes

How are salt and density related?
Bodies of water form layers based on differences in density. How does salt affect density?

1. Read and complete a lab safety form.
2. Half-fill a **glass** with **water.**
3. Carefully place a **hard-cooked egg** in the water. Observe what happens. Remove the egg.
4. Add 5–10 tablespoons of **salt** and stir until all the salt is dissolved.
5. Place a **ladle** or a **spoon** inside the glass and slowly pour tap water over it until the glass is three-fourths full. Gently remove the ladle or the spoon. Be careful not to disturb the layer of salt water.
6. Gently place the egg in the glass and observe.

Think About This
1. Explain any differences that you observed.
2. 🔑 **Key Concept** Do you think it is easier to float in the ocean or in a freshwater lake?

Earth's Oceans

Aside from being called the water planet, did you know that sometimes Earth is also called the blue planet? If you have ever seen a photograph of Earth taken from space, such as the one in **Figure 1,** you know that Earth appears mostly blue. Earth appears blue because water covers 70 percent of its surface. Most of Earth's water—97 percent—is salt water in the oceans.

Earth's oceans are all connected. However, scientists separate the oceans into five main bodies:

- The Pacific Ocean is the largest and deepest ocean. It is larger than all of Earth's combined land area.

- The Atlantic Ocean is half the size of the Pacific. It occupies about 20 percent of Earth's surface.

- The Indian Ocean is between Africa, India, and the Indonesian Islands. It is the third largest ocean.

- The Southern Ocean surrounds Antarctica. It is Earth's fourth largest ocean. Ice covers some of its surface all year.

- The Arctic Ocean is near the North Pole. It is the smallest and shallowest ocean. Ice covers some of its surface all year.

In this lesson, you will read about the formation of the oceans, their physical and chemical characteristics, and the importance of the oceans' natural resources.

Figure 1 Earth appears blue from space because its water reflects blue wavelengths of light.

Lesson 1
EXPLORE

Figure 2 Volcanic eruptions on Earth today add water vapor to the atmosphere, just as they did billions of years ago.

Formation of the Oceans

Evidence indicates that Earth's oceans began to form as early as 4.2 billion years ago (bya). That is only a few hundred million years after Earth formed. Earth was very hot and active when it was young. Many volcanoes covered its surface. Like the volcano shown in **Figure 2,** these ancient volcanoes erupted huge amounts of gas. Much of the gas was made of water vapor, with small amounts of carbon dioxide and other gases. Over time, these gases formed early Earth's atmosphere.

Condensation As water moves through the water cycle, illustrated in **Figure 3,** water vapor in the atmosphere cools and condenses into a liquid. Tiny droplets of liquid combine and form clouds. As early Earth cooled, the water vapor in its atmosphere condensed and precipitated. Rain fell for tens of thousands of years, collecting on Earth's surface in low-lying basins. Eventually, these basins became the oceans.

Asteroids and Comets Evidence suggests a second source of water for Earth's oceans. During the time when oceans formed, many icy comets and asteroids from space collided with Earth. The melted ice from these objects added to the water filling Earth's ocean basins.

 Reading Check What are the sources of Earth's oceans?

Tectonic Changes Earth's oceans change over time. As tectonic plates move, new oceans form and old oceans disappear. However, the volume of water in the oceans has remained fairly constant since the first oceans formed.

Figure 3 Earth's water continually evaporates from the ocean and returns to the ocean through the water cycle.

Composition of Seawater

The rain that fell to Earth's surface billions of years ago washed over rocks and dissolved minerals. The minerals contained substances that formed salts. Rivers and streams carried these substances to ocean basins. Some substances also came from gases released by underwater volcanoes. Together, these substances made the water salty, as shown in **Figure 4.**

 Key Concept Check Why is seawater salty?

Salinity *is a measure of the mass of dissolved solids in a mass of water.* Salinity is usually expressed in parts per thousand (ppt). For example, **seawater** *is water from a sea or ocean that has an average salinity of 35 ppt.* This means that if you measured 1,000 g of seawater, 35 g would be salts and 965 g would be pure water.

The salinity of seawater changes in areas where rivers enter the ocean, such as in an estuary. There, seawater becomes brackish. **Brackish** *water, or brack water, is freshwater mixed with seawater.* The salinity of brackish water is often between 1 ppt and 17 ppt.

Substances in Seawater

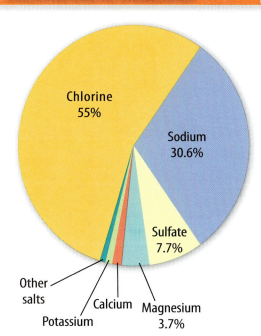

Figure 4 Five elements and one compound account for 99 percent of the dissolved substances in seawater. Evidence suggests that the percentages shown in this circle graph have been fairly consistent for millions of years.

Visual Check Sodium makes up what percentage of the dissolved substances in seawater?

Inquiry MiniLab   20 minutes

How does salinity affect the density of water?

Salt water is more dense than freshwater. How much salt do you need to add to freshwater to make it dense enough to float an egg?

1. Read and complete a lab safety form.
2. Fill a **jar** with 1,000 mL of **water.** Carefully add a **hard-cooked egg** to the water. Observe the egg's position.
3. Use a **stirring rod** to stir 20 g of **salt** into the water. Again observe the egg's position.
4. Add salt in increments of 10 g. After each addition, stir the salt into the water and observe the egg. Continue to add salt in 10-g increments until the egg floats.

Analyze and Conclude

1. **Calculate** the salinity of the water in which the egg floated.
2. **Key Concept** How does salinity affect the density of water?

Seafloor Topography

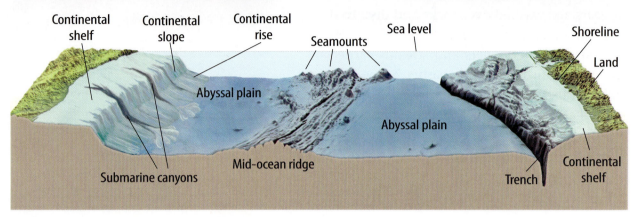

Figure 5 An ocean's seafloor is shaped like a basin. Some features of ocean basins are continental shelves, continental slopes, continental rises, abyssal plains, mid-ocean ridges, seamounts, and trenches.

Visual Check Where is new seafloor created?

WORD ORIGIN
abyssal
from Greek *abyssos*, means "bottomless"

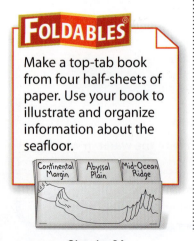

Make a top-tab book from four half-sheets of paper. Use your book to illustrate and organize information about the seafloor.

The Seafloor

What do you think the ocean bottom looks like? You might be surprised to learn that the seafloor has features similar to features on land, such as plains, plateaus, canyons, and mountains.

Continental Margins

The part of an ocean basin next to a continent is called a continental margin. A continental margin extends from a continent's shoreline to the deep ocean. It is divided into three regions, which are illustrated in **Figure 5.** The continental shelf is the shallow part of a continent nearest the shore. The continental slope is the steep slope that extends from the continental shelf to the deep ocean. The continental rise is at the base of the slope. It is where sediments accumulate that fall from the continental slope.

Abyssal Plains

Examine **Figure 5** again. Notice the abyssal plains. **Abyssal plains** *are large, flat areas of the seafloor that extend across the deepest parts of the ocean basins.* Thick layers of sediment cover abyssal plains. In some areas, underwater volcanoes rise from the abyssal plains and form islands that extend above the ocean's surface.

Mid-Ocean Ridges

At places on the seafloor where tectonic plates pull apart, volcanic mountains form. These underwater mountains are called mid-ocean ridges. Mid-ocean ridges form a continuous mountain range that extends through all of Earth's ocean basins. It is Earth's tallest and longest mountain range, measuring more than 65,000 km in length. As the plates slowly move apart at mid-ocean ridges, lava erupts and then cools forming new seafloor.

Ocean Trenches

Earth's oceans have an average depth of about 4,000 m. However, in areas where an oceanic tectonic plate collides with a continental plate, a deep canyon, or trench, forms along the edge of the abyssal plain. A trench is shown in **Figure 5.** Trenches are the deepest parts of the ocean. The Mariana Trench, in the western Pacific Ocean, is more than 11,000 m deep. The bottom of the Mariana Trench extends farther below sea level than Mount Everest is above sea level.

 Key Concept Check Describe some features of the seafloor.

Deep Ocean Technology

Today, scientists use submersibles and other technologies to explore the seafloor. A submersible is an underwater vessel which can withstand extreme pressure at great depths. One famous submersible, DSV *Alvin*, set a deep-ocean record by diving to the bottom of the Mariana Trench.

In the future, remotely operated vehicles (ROVs) are likely to be used more frequently. These unmanned submersibles can be operated from a control center on a ship. Operators can see video images sent back from the ROVs and can control their propellers and **manipulator** arms. ROVs are safer, cheaper, and can generally provide more research data than manned submersibles.

Resources from the Seafloor

The seafloor contains valuable resources. **Table 1** illustrates some of the resources on or beneath the seafloor. There are two main categories of seafloor resources—energy resources and minerals. Energy resources, such as oil, natural gas, and methane hydrates, are beneath the ocean floor on continental margins. Most mineral deposits, such as the manganese nodules shown in **Table 1,** are on abyssal plains. Some minerals, including gold and zinc, have also been discovered at mid-ocean ridges.

Table 1 Resources from the Ocean Floor

Oil and Natural Gas
These deposits are beneath the seafloor on continental margins. Many platforms for oil extraction have been built in the Gulf of Mexico.

Methane Hydrates
Deposits of methane gas in deep-sea sediments are called methane hydrates. They are a potential but as yet unrealized source of energy similar to fossil fuels.

Mineral Deposits
Minerals on the seafloor include manganese nodules. These nodules form when metals precipitate out of seawater. They are potentially valuable, but no large-scale mining exists.

Table 1 Resources found on or below the seafloor include oil, methane hydrates, and manganese nodules.

ACADEMIC VOCABULARY
manipulate
(verb) to operate with hands or by mechanical means in a skillful manner

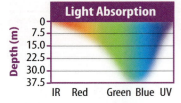

Figure 6 Wavelengths of blue and green light reach deeper into the ocean than those of red, orange, and yellow light.

REVIEW VOCABULARY

photosynthesis
a chemical process in which light energy, water, and carbon dioxide are converted into sugar

Figure 7 The surface zone begins at the ocean surface and reaches a depth of about 200 m. The middle zone begins below the surface zone and reaches a depth of about 1,000 m. The deep zone is below the middle zone.

Zones in the Oceans

Scientists divide oceans into distinct regions, or zones, based on physical characteristics. These characteristics include the amount of sunlight, temperature, salinity, and density.

Amount of Sunlight

If you have ever swum in a lake or an ocean, you might have noticed that the deeper the water, the darker it was. Light from the Sun penetrates below the ocean's surface. However, as depth increases, the wavelengths of light are not absorbed equally. Because of this, some colors penetrate deeper than others, as illustrated by the graph in **Figure 6.**

Surface Zone The area of shallow seawater that receives the greatest amount of sunlight is the surface zone, or sunlit zone. This zone is located above the dashed line shown in **Figure 7.** Most organisms that perform **photosynthesis** live here.

Middle Zone By the time sunlight reaches the middle zone, or twilight zone, most of light's wavelengths have been absorbed. This zone receives only faint blue-green light. The area between the two dashed lines in **Figure 7** represents the middle zone.

Deep Zone Plants do not grow in the deep zone, or midnight zone, where there is no light. Most deep-sea animals, such as the squid shown in **Figure 7,** make their own light in chemical process called bioluminescence (BI oh LEW mah NE cents).

Reading Check Why don't plants grow in the deep zone?

Ocean Layers

Just as oceans have zones of light, they also have zones of temperature, salinity, and density. Notice in **Figure 8** that temperature, salinity, and density vary with depth. Sometimes these characteristics can change abruptly within a relatively short change of depth. Abrupt changes in these characteristics can create distinctive layers of seawater.

 Key Concept Check Why does seawater form layers?

Figure 8 Temperature, salinity, and density vary in the top 1,000 m of Earth's oceans.

Visual Check Below what depth does all ocean water have approximately the same temperature?

Changes in Temperature, Salinity, and Density

Changes in Temperature As shown in the graph to the right, temperature changes abruptly between 250 m and 900 m in temperate and tropical regions (solid line). As depth increases, water in these regions cools rapidly. That is because there is less sunlight to warm water as depth increases.

In contrast, the temperature of polar water (dotted line) remains fairly constant. This is because sunlight intensity at Earth's poles is weaker than it is in temperate and tropical regions. Polar water at all depths is cold.

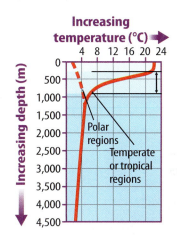

Changes in Salinity The top 500 m of warm water in temperate and tropical regions is saltier than polar water. Warm water evaporates more rapidly than cold water. When water evaporates, salt is left behind; this increases salinity at the surface.

In polar regions, freshwater from melting glaciers decreases the salinity at the surface. However, when ice forms, salt is left behind in the water. The remaining cold, salty water becomes denser and sinks to a deeper layer.

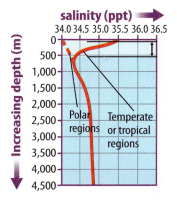

Changes in Density Seawater density is related to temperature and salinity. Cold water is denser than warm water. Salt water is denser than freshwater. Because of density differences, ocean water is layered. The densest layers are on the bottom; the least dense layers are on top.

Notice in the graph to the right that water density in polar regions remains fairly constant. Keep this in mind when you read about density currents in Lesson 3.

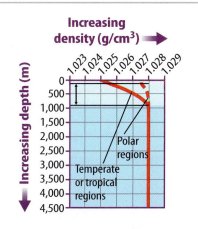

Lesson 1
EXPLAIN

Lesson 1 Review

 Assessment   Online Quiz

Visual Summary

 Condensation of water vapor from volcanic eruptions formed Earth's oceans.

 The seafloor has topographic features such as mountains, plains, and trenches.

 Sunlight, temperature, salinity, and density of seawater change with depth.

FOLDABLES

Use your lesson Foldable to review the lesson. Save your Foldable for the project at the end of the chapter.

What do you think NOW?

You first read the statements below at the beginning of the chapter.

1. Oceans formed about 4 billion years ago.
2. The seafloor is flat.

Did you change your mind about whether you agree or disagree with the statements? Rewrite any false statements to make them true.

Use Vocabulary

1. **Compare** brackish water and seawater.
2. **Use the term** *salinity* in a complete sentence.

Understand Key Concepts

3. Which resource from the oceans is used as a source of energy?
   A. manganese   C. salt
   B. natural gas   D. sand
4. **Explain** why oceans are salty.
5. **Describe** how seawater forms layers.

Interpret Graphics

6. **Organize information** Copy and fill in the graphic organizer below to identify three zones in the ocean based on the amount of light reaching each zone.

7. **Identify** which letters in the figure below represent the continental shelf, the continental slope, and the continental rise. How do these areas differ?

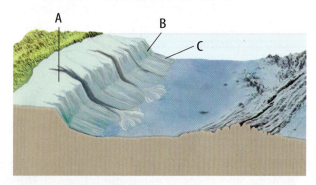

Critical Thinking

8. **Design** Suppose you have been hired to mine manganese nodules off the bottom of the Pacific Ocean. Identify the problems you might have, and design equipment that would enable you to get the nodules to the surface.

Exploring Deep-Sea Vents

CAREERS in SCIENCE

Meet Susan Humphris, a scientist investigating how deep-sea vents affect the chemistry of the oceans.

More than a mile below the ocean's surface is an extraordinary world. Dark as night, nearly freezing, and under crushing pressure, this part of the seafloor is one of the least explored places on Earth. Amazingly, communities of unusual organisms, such as tubeworms and giant clams, thrive here near underwater hot springs called deep-sea vents. These creatures get their energy from chemicals rather than sunlight.

Susan Humphris, a scientist at Woods Hole Oceanographic Institution, studies this unique environment. As a marine geochemist, she investigates the composition, or chemical makeup, of the rocks and seawater around deep-sea vents. When the icy seawater meets hot volcanic rocks deep below the seafloor, they undergo chemical reactions. Humphris investigates how these reactions change the composition of volcanic rocks, seawater, and the ocean as a whole.

To observe the deep-sea vents and collect rock samples, she uses specialized vehicles that can travel to the seafloor. Some vehicles are unmanned and are operated remotely by people on a ship. But in one vehicle, called *Alvin*, Humphris can travel to the bottom of the ocean and explore deep-sea vents herself.

Back in the lab, Humphris analyzes the composition of her volcanic rock samples. By comparing them with other volcanic rock, she can determine which elements were exchanged during the chemical reactions between rocks and seawater. Humphris and other scientists have determined that deep-sea vents have affected all the seawater in the world's oceans.

Deep-Sea Vents

Deep-sea vents form near mid-ocean ridges—long chains of underwater volcanoes that encircle Earth. When icy seawater seeps into deep cracks in Earth's crust, it becomes superheated as it contacts hot volcanic rock. This causes the water to gush upward from the seafloor as a deep-sea vent. The entire ocean circulates through deep-sea vents every 1 million to 10 million years.

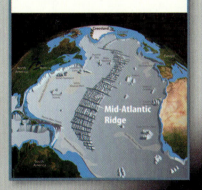

It's Your Turn

JOURNAL ENTRY Imagine you are piloting *Alvin*. Write a journal entry about your expedition to a deep-sea vent. Include descriptions and drawings of what you see as you travel to the dark seafloor.

Lesson 1 EXTEND

Lesson 2

Ocean Waves and Tides

Reading Guide

Key Concepts
ESSENTIAL QUESTIONS
- What causes ocean waves?
- What causes tides?

Vocabulary
tsunami p. 575
sea level p. 576
tide p. 576
tidal range p. 576
spring tide p. 577
neap tide p. 577

 Multilingual eGlossary

Video BrainPOP®

Inquiry Surfing Under a Wave?

Is this surfer confused? Why is he under the wave? What do you think happens to a wave's energy below the surface?

Launch Lab

10 minutes

How is sea level measured?

The ocean surface is changing constantly as a result of waves, tides, and currents. In a matter of seconds, a wave can cause the ocean surface to rise and fall by several meters. In a matter of hours, a tide can also raise or lower the level of the sea by several meters.

1. Read and complete a lab safety form.
2. Half-fill a **clear container** with **water.**
3. Slowly and steadily rock the container back and forth to produce waves.
4. While you gently rock the container, another student should look through the side of the container and mark the peaks and valleys of the waves with a **wax pencil**.
5. Using a **ruler**, measure the difference between the two marks. The midpoint of this measurement is equivalent to sea level.

Think About This

1. How do you think sea level changes when wind speed changes?
2. **Key Concept** How do you think oceanographers determine sea level?

Parts of a Wave

Have you ever been caught in a crashing wave? It might have been hard to catch your breath. Even if you dive deep below a wave, you can still feel some of the wave's energy. The surfer shown on the opposite page is duck diving—ducking beneath a wave to avoid the wave's full power.

There are different kinds and sizes of waves in the oceans, but all waves have the same basic parts. As shown in **Figure 9,** the crest is the highest part of a wave. The trough is the lowest part of a wave. The wave height is the vertical distance between the crest and the trough. The wavelength is the horizontal distance from crest to crest or from trough to trough.

 Reading Check How is wavelength measured?

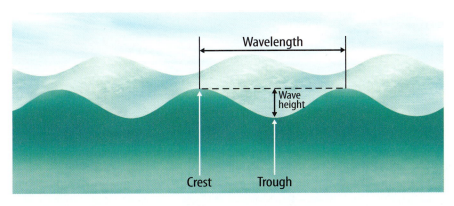

Figure 9 Ocean waves have crests and troughs.

Concepts in Motion
Animation

Lesson 2 **573**
EXPLORE

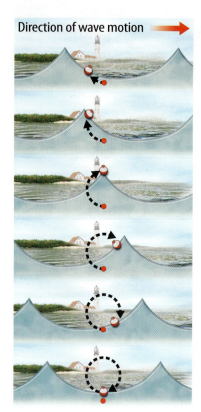

Direction of wave motion →

Figure 10 Just like a fishing bobber, a water particle moves in a circle as a wave passes.

Surface Waves

Wind causes the waves that roll onto a beach. They are often called surface waves. Friction from wind drags across the water's surface, causing it to ripple. The small ripples eventually become larger waves.

 Key Concept Check What causes ocean surface waves?

Surface waves range in size from tiny ripples to huge waves several meters high. Three factors affect the size of surface waves—wind speed, time, and distance. The faster, longer, and farther the wind blows, the larger the resulting waves. For example, some of the largest wind-driven waves form in the Southern Ocean. It experiences fast and continuous winds that blow all the way around Antarctica.

Wave Motion

If you watch a wave wash onto a beach, you might think that a wave transports water from one location to another. However, the motion of a water particle in a wave is circular. After a wave passes, the water particle returns approximately to its original position, as shown in **Figure 10**.

The circular motion of water particles extends below the surface. However, as depth increases, the circular motion decreases. At a certain depth, called the wave base, wave motion stops. This depth is equal to a distance of one–half the wavelength of the wave above it, as illustrated in **Figure 11**.

Wave Motion at Depth

Figure 11 The circular motion of water particles becomes smaller and smaller with depth.

Visual Check If the wavelength of a surface wave is 40 m, how deep would a scuba diver have to go before feeling no wave motion?

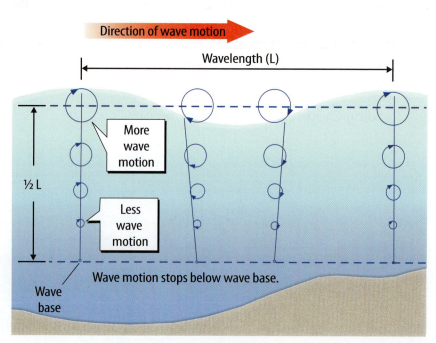

574 Chapter 16 EXPLAIN

When Surface Waves Reach Shore

As a wave moves into shallow water, it changes shape and size. The change begins when the base of the wave comes in contact with the sloping seafloor, as shown in **Figure 12.** As the base of the wave drags on the seafloor, the wave's speed decreases. At the same time, the wavelength shortens and the wave height increases. When the wave reaches a certain height, the wave base can no longer support the crest, and the wave collapses, or breaks. This type of wave is called a breaker. After a wave breaks, the water surges forward onto shore.

Make a shutter-fold book and use it to organize your notes about surface waves and tides.

Breakers

Figure 12 A wave changes shape when its base comes in contact with the seafloor.

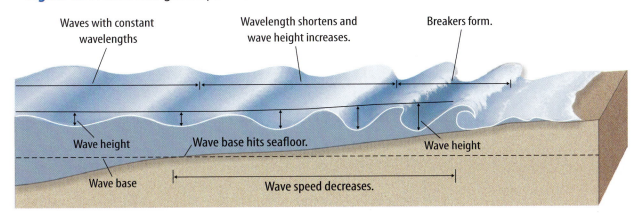

Tsunamis

You might have heard of another type of ocean wave called a tsunami. A **tsunami** (soo NAH mee) *is a wave that forms when an ocean disturbance suddenly moves a large volume of water.* It can be caused by an underwater earthquake or landslide, a volcanic eruption, or even ice breaking away from a glacier.

Key Concept Check What can cause a tsunami?

Far from shore, a tsunami has a short wave height, often less than 30 cm high. However, the wavelength can be hundreds of kilometers long. As a tsunami approaches shore, it slows down and grows higher. Many tsunamis grow only a few meters high as they move onto shore, but some can rise as high as 30 m.

Unlike a common wind-driven wave, the water from a tsunami just keeps coming. As a result, tsunamis can cause much damage. In 2004, a series of tsunamis caused by an underwater earthquake in the Indian Ocean killed more than 225,000 people in 11 countries and destroyed entire villages.

WORD ORIGIN

tsunami
from Japanese *tsu*, means "harbor"; and *nami*, means "wave"

Math Skills

Use Statistics
Find the **mean** by adding the numbers in a data set and dividing by the number of items in the set. The **range** is the difference between the largest and smallest numbers in a set of data.

Example: In a 48-hour period, high tides were measured at 0.701 m, 0.649 m, 0.716 m, and 0.661 m above sea level. What is the range and mean of the high tides?

Range = 0.716 m − 0.649 m
= 0.067 m

Mean = (0.701 m + 0.649 m + 0.716 m + 0.661 m) ÷ 4
= 2.73 m ÷ 4 = 0.682 m

Practice
During the same 48-hour period, the low tides were measured at 0.018 m, 0.103 m, 0.048 m, and 0.091 m below sea level.

a. What is the range of the low tides?

b. What is the mean of the low tides?

 Review
- Math Practice
- Personal Tutor

Tides

When measuring sea level, scientists take into account changes to the ocean's surface caused by waves. **Sea level** is the *average level of the ocean's surface at any given time.* Scientists who measure sea level also take into account changes to the ocean's surface caused by tides. **Tides** *are the periodic rise and fall of the ocean's surface caused by the gravitational force between Earth and the Moon, and between Earth and the Sun.*

The Moon and Tides

The gravitational force that causes the largest tides is between Earth and the Moon. The attraction between them produces two bulges on ocean surfaces—one bulge on the side of Earth facing the Moon and one bulge on the side of Earth facing away from the Moon. The bulges represent high tides. High tide is the highest level of an ocean's surface. Low tide, the lowest level of an ocean, occurs between the two bulges. The difference between high tide and low tide in one coastal area is shown in **Figure 13.**

 Key Concept Check What causes the largest tides?

Topography and Tides

The coastlines of continents, the shape and size of ocean basins, and the depth of the oceans affect tides. The Atlantic coast experiences two alternating high and low tides almost daily. In contrast, the Gulf of Mexico experiences one high tide and one low tide each day.

The size of tides also varies on different areas of Earth's surface. In some areas, the difference between low tide and high tide is as small as 1 m. In other areas, the difference is as great as 15 m. As shown in **Figure 13,** *the difference in water level between a high tide and a low tide is the* **tidal range.**

Figure 13 Tides change the level of the ocean's surface.

Tidal Forces

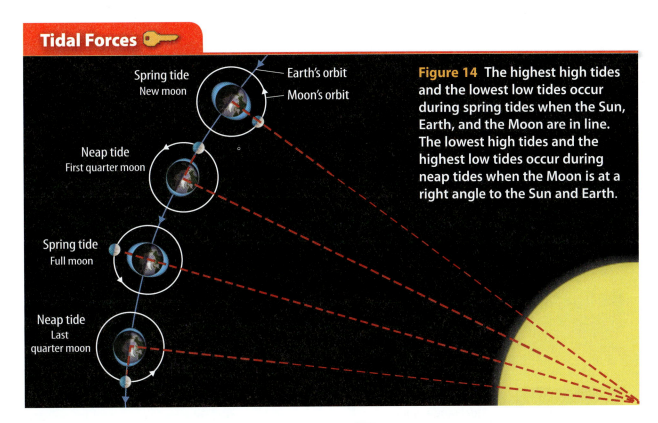

Figure 14 The highest high tides and the lowest low tides occur during spring tides when the Sun, Earth, and the Moon are in line. The lowest high tides and the highest low tides occur during neap tides when the Moon is at a right angle to the Sun and Earth.

Spring Tides

Tidal ranges are not constant. They vary depending on the positions of the Sun and the Moon with respect to Earth. Notice in **Figure 14** that when Earth, the Moon, and the Sun are aligned, the Moon is new or full. The gravitational pull on the oceans is strongest when the two forces act together. As a result, the tidal range is larger than normal. High tides are higher and low tides are lower. A **spring tide** has the greatest tidal range and occurs when Earth, the Moon, and the Sun form a straight line.

Neap Tides

Look at **Figure 14** again. During a first quarter moon and a last quarter moon, the Moon is at a right angle to Earth and the Sun. The gravitational forces between Earth and the Moon and between Earth and the Sun act against each other. This means that high tides are lower than normal while low tides are higher than normal. A **neap tide** has the lowest tidal range and occurs when Earth, the Moon, and the Sun form a right angle.

Reading Check What is a neap tide?

Inquiry MiniLab   20 minutes

Can you analyze tidal data?

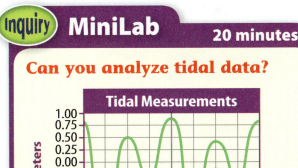

Analyze and Conclude

1. **Determine** how many high tides and low tides there are in a 24-hour period.

2. **Compare** Is the height of the high tides the same within a 24-hour period? What about the height of the low tides?

3. **Calculate** the tidal range between 12 A.M. and 6 A.M. on Day 1.

4. **Key Concept** Suppose the data represent spring tides. How would the tidal data collected during a neap tide be different?

Lesson 2
EXPLAIN

Lesson 2 Review

Visual Summary

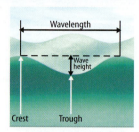

All waves have the same basic features.

Wavelength shortens and wave height increases as a wave nears the shoreline.

Tidal ranges vary from location to location, but can be up to 15 m in difference.

FOLDABLES

Use your lesson Foldable to review the lesson. Save your Foldable for the project at the end of the chapter.

What do you think NOW?

You first read the statements below at the beginning of the chapter.

3. Waves move water particles from one location to another.

4. The wind causes tides.

Did you change your mind about whether you agree or disagree with the statements? Rewrite any false statements to make them true.

Use Vocabulary

1. **Use the term** *tsunami* in a complete sentence.
2. **Define** *tide* in your own words.

Understand Key Concepts

3. **Explain** how the Moon causes tides.
4. **Compare and contrast** the causes of surface waves and tsunamis.

Interpret Graphics

5. **Organize Information** Copy and fill in the graphic organizer below to describe spring tides and neap tides.

	Positions of Earth, Moon, and Sun
Spring tides	
Neap tides	

6. **Explain** how the figure below represents the movement of water in a wave.

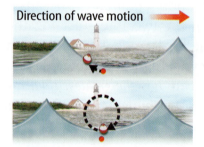

Critical Thinking

7. **Design** an experiment to measure the average tidal range in a coastal area during one month.

Math Skills

8. In a certain location, high tides for one day measure 8.30 m and 8.00 m. The low tides measure 0.500 m and 0.220 m.
   A. What is the range of the tides?
   B. What is the mean low tide?

Inquiry Skill Practice — Analyze Data 20 minutes

High Tides in the Bay of Fundy

The tides in the Bay of Fundy in Eastern Canada have the greatest tidal ranges of any tides on Earth. As a tide enters the Bay of Fundy, it is channeled into an increasingly narrower space. Topography of the land directly affects the tidal range.

The lines on the map of the Bay of Fundy below are similar to contour lines on a topographic map. Tidal height data has been collected along each line and then averaged to determine the mean height of the highest tide at that location across the width of the bay.

Learn It
Analyze the data on the map, to make a graph showing the change in tidal heights from the mouth of the bay to the town of Truro.

Try It
1. Make a data table with three columns in your Science Journal. Label the columns: High Tide (m), Distance from the Mouth of the Bay (cm), Distance from the Mouth of the Bay (m).
2. Use the map scale of the Bay of Fundy below and a metric ruler to determine the distance each high tide is from the mouth of the bay. Convert centimeters on your ruler to meters on the map. Record your information in your data table.
3. Using your data, graph the distance from the mouth of the bay along the x-axis (in m), and tidal height along the y-axis. Give your graph a title.

Apply It
4. **Describe** how the highest tides changed with distance.
5. **Infer** how the tides in the Bay of Fundy might change when Earth, the Moon, and the Sun are in a straight line. How might the tides change when Earth, the Moon, and the Sun are at an angle?
6. **Key Concept** Identify factors that affect tides in the Bay of Fundy.

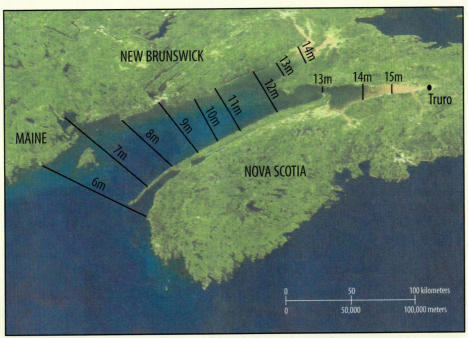

Lesson 3

Reading Guide

Key Concepts 🔑
ESSENTIAL QUESTIONS

- What are the major types of ocean currents?
- How do ocean currents affect weather and climate?

Vocabulary
ocean current p. 581
gyre p. 582
Coriolis effect p. 582
upwelling p. 583

g Multilingual eGlossary

Video BrainPOP®

Ocean Currents

Inquiry Where are they going?

Can you find the Florida Current in this satellite photo? The curve of the clouds gives it away. As the clouds move between Florida and Cuba, they follow the same path as the current. Why do you think clouds and currents sometimes follow the same path?

Inquiry Launch Lab

10 minutes

How does wind move water?

Wind pushes the water near shorelines in different directions. Objects released here go around and around in the waves. What would happen farther out in the ocean?

1. Read and complete a lab safety form.
2. Half fill a **container** with water.
3. Position a **fan** so it can blow across the water's surface.
4. Put two drops of **food coloring** on the surface of the water closest to the fan. Turn the fan to a low setting to produce waves.
5. Observe what happens to the food coloring.

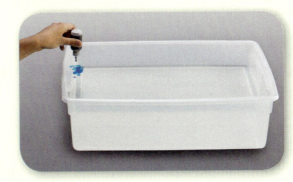

Think About This

1. Explain the movement of the food coloring in your Science Journal.
2. What types of objects do you think the wind can move in the ocean?
3. 🔑 **Key Concept** If you were on a boat about 3 km from shore and threw a rubber ball into the water, what do you think would happen?

Major Ocean Currents

During a storm in 1990, 40,000 pairs of shoes fell off a cargo ship in the middle of the Pacific Ocean. Months later, beachcombers began finding the shoes on the coasts of Oregon and Washington. How did the shoes get there? An ocean current carried them. *An* **ocean current** *is a large volume of water flowing in a certain direction.*

Surface Currents

Recall that wind transfers energy to water and forms waves. Wind also transfers energy to water and forms currents. The friction generated by wind on water can move the water. As wind blows over water, the moving air particles drag on the surface and cause the water to move, just as they drag the wind surfer in **Figure 15**. Wind-driven currents are called surface currents.

Surface currents carry warm or cold water horizontally across the ocean's surface. They extend to about 400 m below the surface and can move as fast as 100 km/day. Earth's major wind belts, called prevailing winds, influence the formation of ocean currents and the direction they move. For example, the trade winds that blow from Africa move warm, equatorial water toward North America and South America.

Figure 15 Just as wind drags this wind surfer across the ocean's surface, the wind also drags the top layer of water across the ocean's surface.

🔑 **Key Concept Check** How do surface currents form?

Lesson 3
EXPLORE
581

Major Ocean Gyres 🔑

▲ **Figure 16** Gyres form on the surface of Earth's oceans.

WORD ORIGIN

gyre
from Latin *gyrus*, means "circle"

Figure 17 The Coriolis effect causes fluids to move clockwise in the Northern Hemisphere and counterclockwise in the Southern Hemisphere. ▼

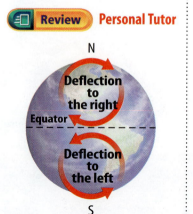

Gyres Earth's oceans contain large, looped systems of surface currents called gyres (JI urz). *A* **gyre** *is a circular system of currents.* As shown in **Figure 16,** the currents within each gyre move in the same direction. However, if you look closely, you can see that the direction of current movement in a gyre is different in each hemisphere. Gyres in the northern hemisphere circle clockwise. Gyres in the southern hemisphere circle counterclockwise.

Coriolis Effect Why do gyres move in different directions? Directions differ because of the Coriolis effect. *The* **Coriolis effect** *is the movement of wind and water to the right or left that is caused by Earth's rotation.* As shown in **Figure 17,** the Coriolis effect causes fluids, such as air and water, to curve to the right in the Northern Hemisphere, in a clockwise direction. In the Southern Hemisphere, the Coriolis effect causes fluids to curve to the left, in a counterclockwise direction.

✓ **Reading Check** What is the Coriolis effect?

Topography The shapes of continents and other landmasses affect the direction and speed of currents. For example, gyres form small or large loops and move at different speeds depending on the land masses they contact. The Florida Current, shown in the photo at the beginning of this lesson, narrows and increases in speed as it passes through the straits of Florida.

Upwelling

Surface currents move water horizontally across the ocean's surface. Not all currents move in a horizontal direction. Some currents move water vertically. **Upwelling** is *the vertical movement of water toward the ocean's surface.* Upwelling occurs when wind blows across the ocean's surface and pushes water away from an area. Deeper, colder water is then forced to the surface. Upwelling often occurs along coastlines. **Figure 18** illustrates how upwelling occurs along the South American coast.

Upwelling brings cold, nutrient-rich water from deep in the ocean to the ocean's surface. This water supports large populations of algae, fish, and other ocean organisms.

 Key Concept Check How does upwelling occur?

Density Currents

Another type of vertical current is a density current. Density currents move water downward. They carry water from the surface to deeper parts of the ocean. Density currents are not caused by wind. They are caused by changes in density.

As you read in Lesson 1, cold water is denser than warm water, and salty water is denser than freshwater. As a surface current moves toward a polar area, the water cools. When seawater freezes, salt is left behind in the surrounding water. Eventually, the cold, salty water becomes so dense that it sinks, as shown in **Figure 19.** Upwelling later brings the current back to the surface. Density currents are important components of ocean circulation. They circulate thermal energy, nutrients, and gases.

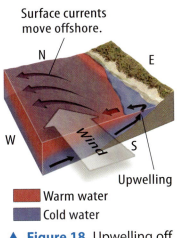

▲ **Figure 18** Upwelling off the South American coast causes cold, deep water to replace warmer water on the surface.

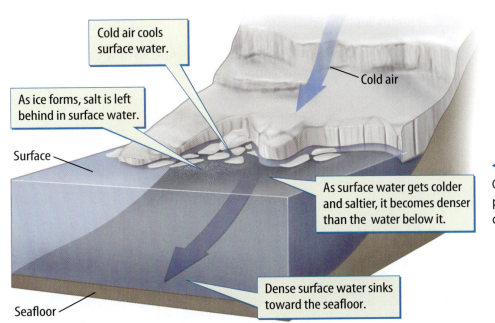

◄ **Figure 19** Cold, salty water sinks, producing a density current.

Figure 20 Higher temperatures are shown in red and yellow. Lower temperatures are shown in green and blue. ▶

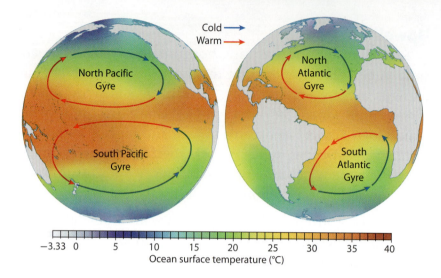

Impacts on Weather and Climate

Solar energy drives convection in the oceans causing warm- and cold-water currents in the gyres shown in **Figure 20.** These two types of surface currents affect weather and climate in different ways. Regions near warm-water currents are often warmer and wetter than regions near cold-water currents. Let's look at some examples.

Surface Currents Affecting the United States

Several warm-water currents affect coastal areas of the southeastern United States. For example, the Gulf Stream, shown in **Figure 21,** transfers lots of thermal energy and moisture to the surrounding air. As a result, summer evenings are often warm and humid. An evening rain is common in these areas.

The cold California Current, also shown in **Figure 21,** affects coastal areas of the southwestern United States. A summer evening along the California coast is often cooler and drier than a summer evening in Florida. Why? This cold-water current releases less thermal energy and moisture to the air.

 Key Concept Check Give an example of how ocean currents can affect weather and climate.

FOLDABLES

Make a shutter-fold book. Use it to record the location of major warm-water currents and cold-water currents and to summarize how they affect weather and climate.

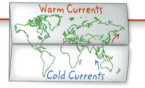

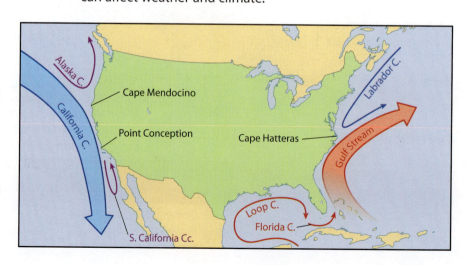

Figure 21 The Gulf Stream is a warm-water current. The California Current is a cold-water current.

Visual Check Hypothesize why hurricanes might be more common in the eastern US than in the western US. ▶

Global Conveyor Belt

Figure 22 A global belt of surface currents and density currents distributes thermal energy on Earth.

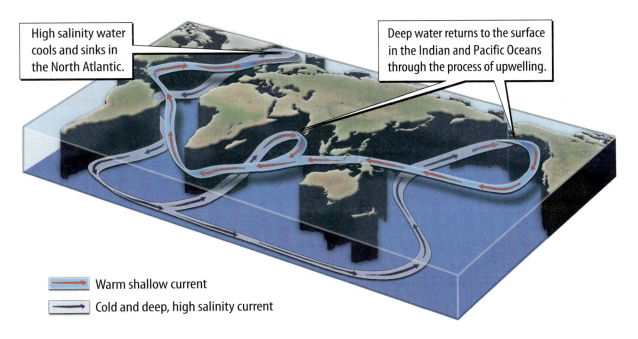

The Great Ocean Conveyor Belt

Aside from gyres, there is another large system of ocean currents that affects weather and climate. This current system is called the Great Ocean Conveyor Belt, illustrated in **Figure 22**. Scientists use this model to explain how ocean currents circulate thermal energy around Earth.

In this model, density currents in the North Atlantic Ocean and the Southern Ocean "run" the conveyor belt. Water in those regions is so cold and dense that it sinks to the ocean bottom and travels along the seafloor. Upwellings in the Pacific Ocean and Indian Ocean eventually bring this deep, cold water to the surface where it is warmed by the Sun.

As warm, surface water travels from the equator toward the poles, it releases thermal energy to the atmosphere, which warms the surrounding region. Then, the cold water sinks until it is upwelled at a different location and the cycle repeats. Scientists estimate that it takes about 1,000 years to complete a cycle.

 Key Concept Check How does the Great Ocean Conveyor Belt affect climate?

Inquiry MiniLab 15 minutes

How does temperature affect ocean currents?

1. Read and complete a lab safety form.
2. Fill one **foam cup** with **hot water** and one cup with **ice water.**
3. Place a **glass dish** on top of the cups. Use two other cups for balance, as shown. Half fill the dish with **room-temperature water.**
4. Put two drops of **food coloring** in the dish, one above each water-filled cup. Use one color for cold water, another for hot water. Observe for 10 min.

Analyze and Conclude

1. **Draw** a diagram of your observations in your Science Journal. Label the hot and cold areas in your drawing.
2. **Key Concept Explain** how your observations of the colored water resemble ocean currents.

Lesson 3 Review

Assessment — Online Quiz

Visual Summary

A gyre is a circular system of surface currents.

Density currents move cold water from the ocean surface to deeper parts of the ocean.

A system of surface currents and density currents distributes thermal energy around Earth.

FOLDABLES

Use your lesson Foldable to review the lesson. Save your Foldable for the project at the end of the chapter.

What do you think NOW?

You first read the statements below at the beginning of the chapter.

5. Ocean currents occur on the surface and below the surface.
6. Ocean currents affect climate and weather.

Did you change your mind about whether you agree or disagree with the statements? Rewrite any false statements to make them true.

Use Vocabulary

1. **Use the term** *Coriolis effect* in a complete sentence.
2. A(n) _____ moves water vertically.

Understand Key Concepts

3. What causes a surface current?
   A. Earth's orbit   C. temperature
   B. Earth's rotation   D. wind

4. **Explain** how energy transfers between currents and the atmosphere affect climate.

5. **Illustrate** how upwelling occurs off the coast of California as wind blows from north to south.

Interpret Graphics

6. **Explain** how the surface currents in the figure below affect the western and eastern coasts of the United States.

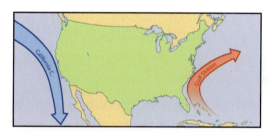

7. **Compare and Contrast** Copy and fill in the graphic organizer below to compare and contrast surface currents and density currents.

	Similarities	Differences
Surface currents		
Density currents		

Critical Thinking

8. **Design** an experiment to show how waves and currents move water in different ways.

9. **Infer** why major fishing grounds are along coastlines.

Inquiry Skill Practice: Interpret Data

30 minutes

How do oceanographers study ocean currents?

Materials

world map

Cargo spills can help oceanographers study ocean currents. The longitude and latitude positions of items from spills that wash ashore contain clues about the direction and speed of currents. Interpret the data below to find out what happened to a cargo of rubber bath toys lost in a January 1992 storm in the North Pacific.

Learn It

Can you make sense of the data in the table at right? You need to **interpret data** before you can draw conclusions about them. Interpret the longitude and latitude positions of toys that washed ashore by marking them on a map.

Try It

1. Mark the longitude and latitude positions on a world map. The other data represent locations where individual bath toys were found. The first data point represents the location of the cargo spill. Label each point with a date.

2. Connect the dots in order of time. Ocean currents don't follow straight lines, so use curved lines. The toys could not float over land, so all the lines you draw should only cross water.

3. Compare the path of the toys to a world map of ocean currents and gyres.

Apply It

4. **Describe** how this data could help oceanographers chart ocean currents.

5. **Hypothesize** how toys traveled to the Atlantic Ocean.

Found Toys		
Date	Latitude	Longitude
January 1992	45°N	178°E
March 1992	44°N	165°W
July 1992	49°N	155°W
October 1992	52°N	135°W
January 1993	59°N	149°W
March 1993	56°N	157°W
July 1993	57°N	170°W
October 1993	59°N	180°E
January 1994	56°N	166°E
March 1994	45°N	155°E
July 1994	47°N	172°E
October 1994	50°N	165°W
January 1995	47°N	140°W
October 2000	46°N	50°W
December 2003	57°N	07°W

6. **Key Concept** What types of ocean currents carry cargo debris around the world?

Lesson 4

Environmental Impacts on Oceans

Reading Guide

Key Concepts
ESSENTIAL QUESTIONS
- How does pollution affect marine organisms?
- How does global climate change affect marine ecosystems?
- Why is it important to keep oceans healthy?

Vocabulary
marine p. 590
harmful algal bloom p. 591
coral bleaching p. 592

 Multilingual eGlossary

 Video

What's Science Got to do With It?

Inquiry Orange Ocean?

The orange-red color of the water in this photograph comes from algae. The algae have formed a huge mat, called an algal bloom, on the ocean's surface. Algal blooms can be beautiful, but some algal blooms harm ocean ecosystems.

Inquiry Launch Lab

15 minutes

What happens to litter in the oceans?

Imagine you are on a boat hundreds of kilometers from shore. You look down at the water and see a sea turtle entangled in plastic. How did this happen?

1. Read and complete a lab safety form.
2. Half-fill a large **bowl** with **water.**
3. Sprinkle **objects** your teacher has supplied into the water.
4. Gently swirl the water in the bowl until the water moves at a constant speed. Try not to create a whirlpool.

Think About This

1. What happened to the objects you sprinkled into the bowl?
2. What do you think happens to litter that is dumped into the ocean?
3. 🔑 **Key Concept** What do you think you can do to prevent ocean pollution?

Ocean Pollution

Have you ever seen a photograph of a shorebird or seal covered in oil? Spills from oil tankers harm wildlife. They also harm the ocean. Any harm to the physical, chemical, or biological health of the ocean ecosystem is ocean pollution. Sometimes ocean pollution comes from a natural source, such as a volcanic eruption. More often, human activities cause ocean pollution.

Sources of Ocean Pollution

Like pollution on land, ocean pollution comes from both point sources and nonpoint sources. Point-source pollution can be traced to a specific source, such as a drainpipe or an oil spill. Nonpoint-source pollution cannot be traced to a specific source. Sewage runoff from land is an example.

Figure 23 shows the proportion of different sources of ocean pollution caused by humans. Notice that only 13 percent of this pollution comes from shipping or offshore mining activity. The rest comes from land. Land-based pollution includes garbage, hazardous chemicals, and fertilizers. Airborne pollution that originates on land, such as emissions from power plants or cars, is also included in this category. So is trash dumped directly into the oceans.

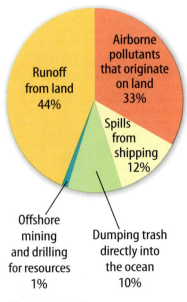

Figure 23 Most ocean pollution caused by humans originates on land.

Lesson 4
EXPLORE
589

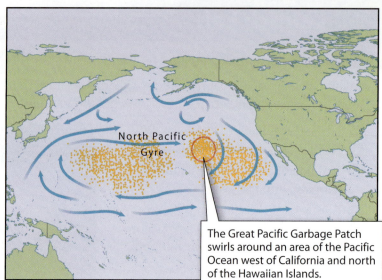

The Great Pacific Garbage Patch swirls around an area of the Pacific Ocean west of California and north of the Hawaiian Islands.

▲ **Figure 24** The North Pacific Gyre traps garbage in areas colored orange on the map. A scientist holds a sample of polluted water from one affected area.

Effects of Ocean Pollution

Ocean pollution has both immediate effects and long-term effects on **marine** ecosystems. Marine *refers to anything related to the oceans.* Chemical waste can be poisonous to marine organisms. Fish and other organisms absorb the poison and pass it up the food chain. A large oil spill can harm marine life. So can solid waste, excess sediments, and excess nutrients.

Solid Waste Trash, including plastic bottles and bags, glass, and foam containers, cause problems for marine organisms. Many birds, fish, and other animals become entangled in plastic or mistake it for food. Plastic breaks up into small pieces but it does not degrade easily. Some of it becomes trapped in the circular currents of gyres. The North Pacific Gyre has collected so much plastic and other debris that some people have named a portion of it "the Great Pacific Garbage Patch." A map showing its location is shown in **Figure 24.** The *Great Pacific Garbage Patch* within the circled area is thought to be twice the size of Texas.

WORD ORIGIN
marine
from Latin *marinus,* means "of the sea"

Figure 25 This satellite image shows sediment from orange-colored soil washing into the ocean. ▼

Excess Sediments Large amounts of land-based sediment wash into oceans, as shown in **Figure 25.** Erosion often occurs on steep coastal slopes after heavy rains. Some of this erosion is natural. But some is caused by humans, who cut down trees near rivers and ocean shorelines. Without the roots of trees and other vegetation to hold sediments in place, the sediments more readily erode. Excess sediments can clog the filtering structures of marine filter feeders, such as clams and sponges. Excess sediments can also block light from reaching its normal depth. Organisms that use light for photosynthesis die.

 Key Concept Check How can excess sediments in oceans affect marine organisms?

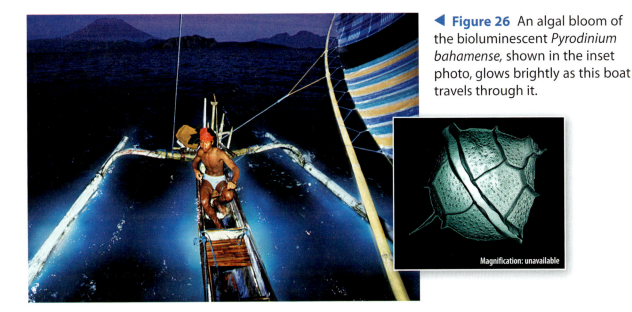

◀ **Figure 26** An algal bloom of the bioluminescent *Pyrodinium bahamense,* shown in the inset photo, glows brightly as this boat travels through it.

Excess Nutrients Algae need nutrients such as nitrogen and phosphorus to survive and grow. However, too many nutrients can cause an explosion in algal populations. An algal **bloom** occurs when algae grow and reproduce in large numbers. The photo at the beginning of this lesson shows how an algal bloom can cause water to turn orange. Algal blooms can also cause water to appear red, green, brown, or even glow at night, as shown in **Figure 26.**

Nitrates and phosphates can be abundant in agricultural runoff as well as coastal upwelling zones. Many scientists suspect that a major source of excess nitrates and phosphates is from land-based fertilizers that wash into oceans.

SCIENCE USE V. COMMON USE

bloom
Science Use a large growth of algae

Common Use a flower

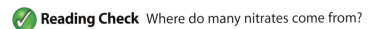 **Reading Check** Where do many nitrates come from?

Many algal blooms are harmless, but others can disrupt marine ecosystems and harm organisms. *A* **harmful algal bloom** *is a rapid growth of algae that harms organisms.* Harmful algal blooms have become more common in recent decades.

Why are some algal blooms harmful? The algae in some algal blooms produce poisonous substances that can kill organisms that eat them. Other algal blooms are so large that they they use up oxygen (O_2) in the water. This can happen when large numbers of algae die and decompose. Decomposition requires O_2. When many algae decompose at the same time, O_2 levels in the water drop. Fish and other marine organisms cannot get enough O_2 to survive. A fish kill resulting from a harmful algal bloom is shown in **Figure 27.**

Figure 27 Excess nitrates that wash into oceans can cause harmful algal blooms which kill fish. ▼

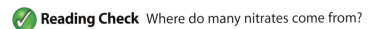 **Key Concept Check** How can excess nutrients in seawater harm fish?

Oceans and Global Climate Change

Solid waste, excess sediments, and algal blooms can cause immediate harm to ocean ecosystems. Other threats to oceans are related to long-term changes in Earth's climate. Climate data indicates that Earth's average surface temperature has increased over the past century. The amount of carbon dioxide (CO_2) in Earth's atmosphere has also increased.

Effects of Increasing Temperature

The increase in Earth's surface temperature has affected oceans in many ways.

Coral Bleaching Some marine organisms, such as coral, are very sensitive to temperature changes. A temperature increase as small as 1°C can cause corals to die, as shown in **Figure 28**. **Coral bleaching** *is the loss of color in corals that occurs when stressed corals expel the algae that live in them.* Coral bleaching harms corals around the world, as shown in **Figure 29**. Coral reefs provide habitat for fish and many other organisms.

Key Concept Check How does water temperature affect corals?

Sea Level As Earth warms, its glaciers and ice sheets melt. This adds water to the oceans and increases sea level. Rising sea levels threaten coastal communities and marine habitats.

Dissolved O_2 The temperature of seawater affects the amount of O_2 dissolved in it. The warmer the water, the less O_2 it contains. Marine organisms need O_2 to survive. As water warms, less O_2 is available, and organisms can die.

Figure 28 Corals contain colorful algae, which provide food for the coral. Without algae, the corals die and appear bleached.

Coral Bleaching

Figure 29 Coral bleaching occurs in many locations around the world.

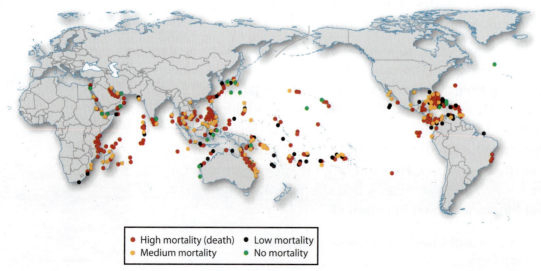

- High mortality (death)
- Medium mortality
- Low mortality
- No mortality

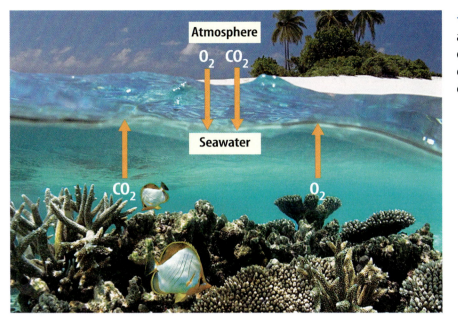

◀ **Figure 30** CO_2 and O_2 are exchanged at the ocean's surface. Waves and currents mix the gases into deeper water.

Effects of Increasing Carbon Dioxide

As illustrated in **Figure 30,** O_2 and CO_2 gases move freely between the atmosphere and seawater. As the amount of CO_2 increases in the atmosphere, the amount of CO_2 dissolved in seawater also increases. This is because of gas exchange at the ocean's surface. These gases dissolve in seawater. Wave action helps mix these gases deeper below the water surface.

CO_2 and pH When CO_2 mixes with seawater, a weak acid called carbonic acid forms. Carbonic acid lowers the pH of the water, making it slightly acidic. Data from recent studies show that the acidity of seawater has increased over the past 300 years. Scientists predict that by 2100, the oceans will become even more acidic, as illustrated in **Figure 31.**

FOLDABLES

Make a chart with three columns and three rows. Label it as shown. Use it to organize information about common gases found in seawater.

 Reading Check Why are oceans becoming more acidic?

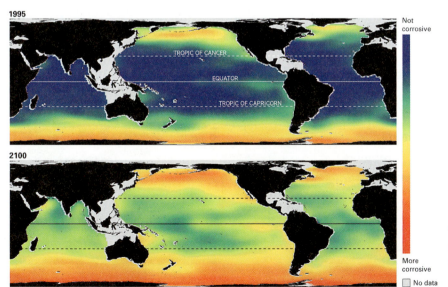

◀ **Figure 31** Scientists predict that oceans in the future will be much more acidic than they are today.

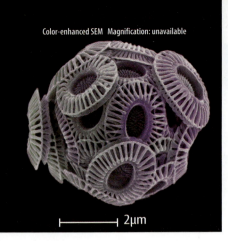

Figure 32 This tiny organism is surrounded by calcium carbonate plates.

Acidity and Marine Life Many marine organisms build shells and skeletons from calcium absorbed from seawater. Snails absorb calcium and make shells. Corals absorb calcium and build reefs. Some algae, like the one shown in **Figure 32,** make protective plates from calcium. As seawater becomes more acidic, it is harder for these organisms to absorb calcium. Increased acidity can cause shells and skeletons to weaken or dissolve. Over time, this could affect food webs. For example, if algae were unable to make protective plates, they would die. Algae form the base of food chains in many marine ecosystems.

Keeping Oceans Healthy

Earth's oceans affect Earth in many ways. As part of the water cycle, they distribute moisture. Ocean currents distribute thermal energy. Oceans provide habitat for algae and other marine organisms. Marine algae release during photosynthesis as much as 50 percent of the O_2 in Earth's atmosphere. Oceans also provide mineral and energy resources. They are a major source of food and income for humans. Keeping oceans healthy is important for the well-being of humans and other organisms on Earth.

Key Concept Check Why is it important to keep oceans healthy?

Inquiry MiniLab
20 minutes

How does the pH of seawater affect marine organisms?

How does increasing acidity affect calcium-containing shells?

1. Read and complete a lab safety form.
2. Copy the table below into your Science Journal.
3. Examine a piece of **brown eggshell** and describe its properties.
4. Place the eggshell in a **plastic cup.**
5. Half fill the cup with **white vinegar.**
6. After 15 minutes, use **forceps** to remove the eggshell. Describe its properties in your Science Journal.

2. **Key Concept** How might long-term effects of increased CO_2 in seawater affect calcium-containing shells and skeletons of marine organisms?

Calcium-containing shells		
Property	Description Before Treatment	Description After Treatment
Hardness		
Thickness		
Appearance		

Analyze and Conclude

1. **Describe** how the eggshell changed.

Lesson 4 Review

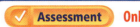 **Assessment** **Online Quiz**

Visual Summary

A harmful algal bloom can cause fish kills.

Increased ocean temperature causes corals to bleach.

Global climate change affects the ocean's chemistry.

FOLDABLES

Use your lesson Foldable to review the lesson. Save your Foldable for the project at the end of the chapter.

What do you think **NOW?**

You first read the statements below at the beginning of the chapter.

7. Most pollution in the oceans originates on land.

8. Global climate change has no effect on marine organisms.

Did you change your mind about whether you agree or disagree with the statements? Rewrite any false statements to make them true.

Use Vocabulary

1 **Define** *harmful algal bloom* in your own words.

2 **Use the term** *marine* in a complete sentence.

Understand Key Concepts

3 How can an increase in CO_2 in the atmosphere affect seawater?
   A. O_2 levels rise
   B. O_2 levels decrease
   C. pH rises
   D. pH decreases

4 **Identify** how excess sediments affect filter feeders.

5 **Construct** a flow chart that shows the steps leading to an algal-bloom fish kill.

Interpret Graphics

6 **Determine Cause and Effect** Copy and fill in the graphic organizer below to list the causes and effects of a decreased pH of seawater.

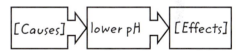

7 **Explain** how environmental conditions can affect the exchange of gases shown in the figure below.

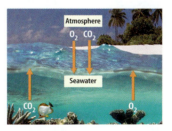

Critical Thinking

8 **Design an experiment** to test the hypothesis that coral bleaching is caused by an increase in water temperature.

9 **Predict** the effect that increased carbon dioxide in the atmosphere could have on food webs in the ocean.

Lesson 4
EVALUATE
595

Inquiry Lab

45 minutes

Predicting Whale Sightings Based on Upwelling

Materials

colored pencils

You are a guide on a blue whale tour departing from Monterey, CA. You have read that upwelling is the vertical movement of cold, nutrient-rich water from the deep ocean to the surface of the ocean. You know upwelling fertilizes the surface of the ocean and creates feeding grounds for fish and plankton-eating whales. Use oceanographic data from satellites and moorings to plan a tour to best view blue whales.

Ask a Question

Where and when can you best observe blue whales near Monterey, CA?

Make Observations

1. Analyze a map of sea surface temperatures (SST) around Monterey Bay.
2. Convert your map to a color contour map. First, construct a legend using colors. Assign warmer pencil colors to warmer temperatures and cooler pencil colors to colder temperatures. Then, outline areas that have the same temperatures in pencil. Finally, color in the sections of the map according to your legend.
3. Study your map noting the position of the upwelling in your Science Journal.
4. Examine the mooring data to the right. Plot both sea surface temperature and wind speed versus day on the same graph. Be sure to label the two vertical axes to reflect the different measurements.

| Data of Wind Direction and Speed ||||
Date	SST (°C)	Wind Direction	Wind Speed (m/s)
23-May	10	N	3
25-May	10	N	8
27-May	9	N	10
29-May	9	N	8
31-May	9	N	4
2-Jun	10	S	−1
4-Jun	12	S	−4
6-Jun	13	S	−3
8-Jun	12	N	7
10-Jun	11	N	5
12-Jun	10	N	8
14-Jun	10	N	7
16-Jun	10	N	7
18-Jun	9	N	9
20-Jun	9	N	11
22-Jun	11	N	4
24-Jun	12	S	−4
26-Jun	13	S	−6
28-Jun	13	-	0
30-Jun	14	S	−1
2-Jul	13	N	6
4-Jul	11	N	9
6-Jul	9	N	10
8-Jul	9	N	10

Chapter 16 EXTEND

5. Analyze your graph and determine under what wind conditions upwelling occurs.

Form a Hypothesis

6. Use your observations of the upwelling to form a hypothesis that gives the location (latitude and longitude) and wind conditions where you could best observe blue whales if you leave on a tour from Monterey, CA.

Test your Hypotheses

7. Use a map showing sightings of blue whales in Monterey Bay to compare your hypothesis to the actual locations where blue whales have been frequently observed. If your hypothesis was not supported, repeat steps 2–3.

8. Compare your prediction of wind conditions for which you could best observe blue whales with another student in your class. If you do not agree, repeat steps 5–6.

Analyze and Conclude

9. **Describe** the location and shape of the upwelling in Monterey Bay.

10. **Analyze** in which direction the wind was blowing when the satellite measurement of sea surface temperature was taken. Explain why this is important to your hypothesis.

11. **Design** a graphic organizer to show the effects of currents, sea surface temperatures, and wind direction on whale feeding areas.

12. **The Big Idea** Explain how currents affect sea life in Monterey Bay.

Communicate Your Results

Design a brochure for a whale watching company based in Monterey, CA. Describe the technology and oceanography that you will use to ensure that your clients observe blue whales.

Inquiry Extension

During the Great Depression, Monterey Bay was one of the largest sardine fisheries in the world. John Steinbeck wrote about the time period in his book *Cannery Row*. Investigate what happened to the sardine fisheries in Monterey Bay during the nineteenth century. Write a Moment in History news report explaining the environmental factors that impacted the growth and decline of the fishery.

Lab Tips

☑ When plotting your data, be sure to use the vertical axis that goes with the data you are plotting.

☑ Draw a line to connect your plot points.

☑ Use two different colors for wind speed and sea surface temperature.

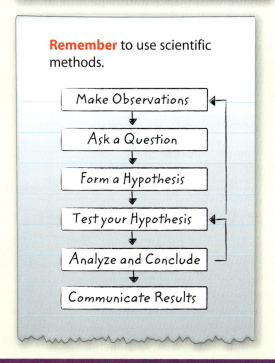

Remember to use scientific methods.

Make Observations → Ask a Question → Form a Hypothesis → Test your Hypothesis → Analyze and Conclude → Communicate Results

Chapter 16 Study Guide

 Oceans affect Earth's climate and weather. They provide resources and habitats. But oceans are threatened by pollution and global climate change.

Key Concepts Summary	Vocabulary
Lesson 1: Composition and Structure of Earth's Oceans  • The salt in the oceans comes mostly from the erosion of rocks and soil. • The seafloor has mountains, deep trenches, and flat plains. • The oceans have zones based on light, temperature, salinity, and density.	**salinity** p. 565 **seawater** p. 565 **brackish** p. 565 **abyssal plain** p. 566
Lesson 2: Ocean Waves and Tides  • The motion of water particles in a wave is circular. • Wind causes most ocean waves, but underwater disturbances cause most **tsunamis.** • The gravitational attraction between Earth and the Moon, and between Earth and the Sun causes **tides.**	**tsunami** p. 575 **sea level** p. 576 **tide** p. 576 **tidal range** p. 576 **spring tide** p. 577 **neap tide** p. 577
Lesson 3: Ocean Currents  • Surface currents, **upwelling,** and density currents are the major **ocean currents.** • Ocean currents affect climate and weather by distributing thermal energy and moisture around Earth.	**ocean current** p. 581 **gyre** p. 582 **Coriolis effect** p. 582 **upwelling** p. 583
Lesson 4: Environmental Impacts on Oceans  • Ocean pollution and climate change affect water temperature and ocean pH, harming **marine** organisms. • A healthy ocean is important because it affects weather and climate, contains habitats for marine organisms, and provides energy resources and food for humans.	**marine** p. 590 **harmful algal bloom** p. 591 **coral bleaching** p. 592

Study Guide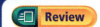

- Personal Tutor
- Vocabulary eGames
- Vocabulary eFlashcards

FOLDABLES Chapter Project

Assemble your lesson Foldables as shown to make a Chapter Project. Use the project to review what you have learned in this chapter.

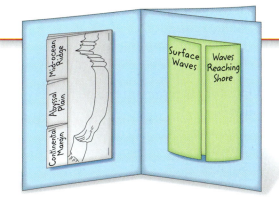

Use Vocabulary

1. Water that has lower salinity than average is _____.

2. Scientists use the term _____ to describe the amount of salt in water.

3. The average height of the ocean's surface is _____.

4. A(n) _____ occurs when Earth, the Moon, and the Sun are in a straight line.

5. A(n) _____ is a large volume of water flowing in a certain direction.

6. A(n) _____ carries warm and cold water in a circular system.

7. A(n) _____ is a vertical movement of water toward the surface.

8. A(n) _____ can occur when increased nutrients cause explosive algal growth.

Link Vocabulary and Key Concepts

Concepts in Motion Interactive Concept Map

Copy this concept map, and then use vocabulary terms from the previous page to complete the concept map.

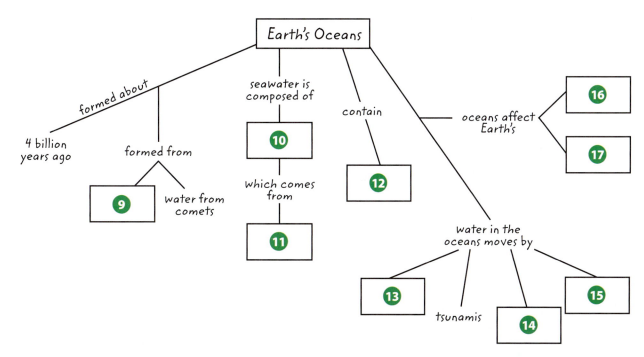

Chapter 16 Review

Understand Key Concepts

1. Based on the circle graph below, which element is most common in seawater?

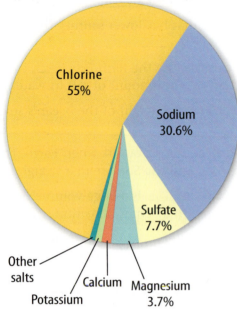

   A. calcium
   B. chlorine
   C. sodium
   D. sulfur

2. Which resource is on abyssal plains?
   A. gravel
   B. manganese nodules
   C. methane hydrates
   D. natural gas

3. Which is NOT a cause of tsunamis?
   A. earthquake
   B. hurricane
   C. landslide
   D. volcanic eruption

4. Which best describes the movement of water in a wave?
   A. circular
   B. horizontal
   C. spiral
   D. vertical

5. Where does an ocean current become most dense?
   A. in polar regions
   B. in temperate regions
   C. near continents
   D. near the equator

6. Which moves water horizontally?
   A. density current
   B. surface current
   C. temperature current
   D. upwelling

7. What does C represent in the figure below?
   A. high tide
   B. low tide
   C. sea level
   D. tidal range

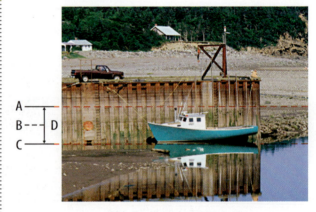

8. Which is one possible effect of an increase in carbon dioxide in the oceans?
   A. Algae grow in excessive amounts.
   B. Corals can't make reefs.
   C. High tides occur more often.
   D. Ocean sedimentation increases.

9. Which is NOT a consequence of rising ocean temperature?
   A. coral bleaching
   B. glacier melting
   C. rising sea level
   D. shells dissolving

Chapter Review

Assessment
Online Test Practice

Critical Thinking

10. **Summarize** the sources of salt in seawater.

11. **Compare** the topography of the ocean floor with the topography of land.

12. **Illustrate** what happens to water particles when a wave passes.

13. **Explain** how a density current might form in the Arctic Ocean.

14. **Design** a model that shows how surface currents form.

15. **Relate** How can cutting trees on land affect life in the ocean?

16. **Assess** the long-term effects of a harmful algal bloom on a marine ecosystem.

17. **Hypothesize** As shown in the figure below, Earth's major warm water currents are on the western boundaries of oceans. Major cold water currents are on the eastern boundaries of oceans. Why are these major currents in these different locations?

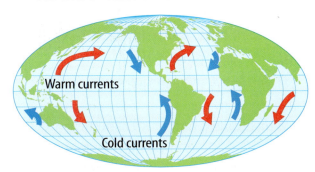

Writing in Science

18. **Compose** a letter to the editor of a newspaper or magazine with ideas of how to reduce human impacts on oceans. Include a main idea, supporting details, examples, and a concluding sentence.

REVIEW THE BIG IDEA

19. Why are oceans important? In what ways are they threatened?

20. How are waves powered? How does movement differ in waves and currents?

Math Skills

Review Math Practice

Use Statistics

Time (Day 1)	Height (m)	Time (Day 2)	Height (m)
00:44	13.1	01:33	13.0
07:13	0.8	08:02	0.9
13:04	13.6	13:54	13.5
19:42	0.3	20:32	0.4

The table above shows the high and low tides during a 48-hour period at the Bay of Fundy. Use the table to answer the questions.

21. What is the range of tides during the 48-hour period.

22. What is the mean of the tides during the 48-hour period?

23. What is the range of the four high tides during the 48-hour period?

24. What is the mean of the four low tides?

Chapter 16 Review • 601

Standardized Test Practice

Record your answers on the answer sheet provided by your teacher or on a sheet of paper.

Multiple Choice

1. Which is a result of increasing acidity of seawater?
   - A Algae populations increase dramatically.
   - B Corals expel algae living within them.
   - C Oxygen is less available to marine organisms.
   - D Shells and skeletons of marine organisms weaken.

2. Which did NOT contribute to the formation of Earth's early oceans?
   - A asteroids
   - B condensation
   - C comets
   - D glaciers

3. What percentage of Earth's water is salt water?
   - A 3%
   - B 55%
   - C 70%
   - D 97%

Use the diagram below to answer question 4.

4. Which seafloor feature does the arrow in the diagram above indicate?
   - A abyssal plain
   - B continental slope
   - C ocean trench
   - D submarine canyon

Use the diagram below to answer question 5.

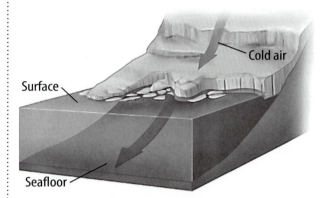

5. Which is formed by the process shown in the diagram above?
   - A gyres
   - B tsunami
   - C density current
   - D surface waves

6. Which results from upwelling in the oceans?
   - A Acidic water dissolves shells.
   - B Cold, dense water sinks.
   - C Marine organisms die.
   - D Surface water gains nutrients.

7. Which causes spring tides and neap tides?
   - A the positions of Earth, the Moon, and the Sun
   - B the rotation of Earth on its axis
   - C the shape of the continental margin
   - D the size and shape of ocean basins

8. As seawater temperature rises, the water contains
   - A less dissolved minerals.
   - B less oxygen.
   - C more coral.
   - D more nutrients.

Standardized Test Practice

Use the diagram below to answer question 9.

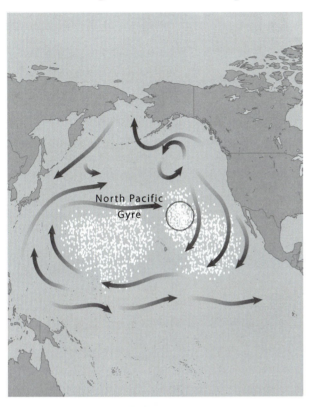

9. The circle on the diagram above indicates a region affected by
   A coral bleaching.
   B frequent tsunamis.
   C excess nitrates and phosphates.
   D pollution from solid waste.

10. Fertilizer runoff from agricultural areas into seawater can cause an excess of
    A acid.
    B carbon dioxide.
    C nutrients.
    D salts.

Constructed Response

Use the diagram below to answer questions 11–13.

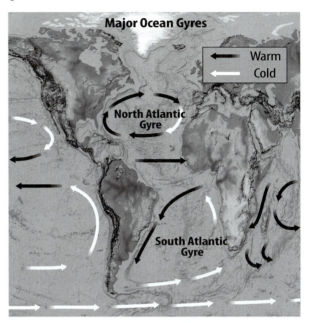

11. What type of current is marked with arrows on the map? How do these currents form? What do they do?

12. Why do these currents move in opposite directions around the North Atlantic and South Atlantic gyres?

13. How do these currents affect the climates of the surrounding continents?

14. What are two ways in which algae benefit other organisms?

15. Why are healthy oceans important to ALL life on Earth?

NEED EXTRA HELP?

If You Missed Question...	1	2	3	4	5	6	7	8	9	10	11	12	13	14	15
Go to Lesson...	4	1	1	1	3	3	2	4	4	4	3	3	3	4	4

Chapter 17

Freshwater

THE BIG IDEA
Where is Earth's freshwater?

Inquiry Why do trees grow here?

Notice the number of trees growing along the Mississippi River. This river ecosystem is green and lush. Rivers are a source of freshwater for many plants and animals. Freshwater helps life sustain itself on Earth.

- What is freshwater?
- How have humans impacted freshwater on Earth?
- Where is most of Earth's freshwater located?

Get Ready to Read

What do you think?
Before you read, decide if you agree or disagree with each of these statements. As you read this chapter, see if you change your mind about any of the statements.

1. On Earth, freshwater occurs only as liquid water.
2. Up to 80 percent of the sunlight that strikes snow or ice is reflected back into space.
3. Organisms living in a lake or a stream are not affected by the amount of oxygen and nutrients in the water.
4. Glaciers can form lake basins.
5. People use groundwater as a source of water.
6. Wetlands can naturally filter pollutants from groundwater.

ConnectED Your one-stop online resource

connectED.mcgraw-hill.com

- Video
- Audio
- Review
- Inquiry
- WebQuest
- Assessment
- Concepts in Motion
- Multilingual eGlossary

Lesson 1

Reading Guide

Key Concepts 🔑
ESSENTIAL QUESTIONS

- How do glaciers affect sea level?
- How does ice and snow cover affect climate?
- How do human activities affect glaciers?

Vocabulary
freshwater p. 607
alpine glacier p. 608
ice sheet p. 609
sea ice p. 611
ice core p. 612

g Multilingual eGlossary

Glaciers and Polar Ice Sheets

Inquiry) Why is the ice melting?

Notice the water streaming off the edge of the iceberg. Why is this ice melting so rapidly? Where does all of the water go? Ice melts as temperature increases. When ice melts, the meltwater eventually enters the oceans, where it can cause a rise in sea level.

Inquiry Launch Lab

10 minutes

Where is all the water on Earth?

Earth is often called the "water planet." That's because about 70 percent of Earth's surface is covered with water stored in the oceans. Where is the rest of Earth's water?

1. Read and complete a lab safety form.
2. Pour 970 mL of water into a **1-L container.** Then, add a drop of **red food coloring.** This represents all of the salt water on Earth.
3. Add 20.7 mL of water to a **clear plastic cup** using a **graduated cylinder**. Then, add a drop of **blue food coloring** to represent all freshwater stored in glaciers.
4. Add 9.0 mL of water to a **clear plastic cup** and then add a drop of **green food coloring.** This represents all the freshwater stored as groundwater.
5. Finally, add one drop (about 0.3 mL) of **yellow food coloring** to a clear plastic cup. This represents all the freshwater in Earth's lakes, rivers, wetlands, atmosphere, and other sources.

Think About This

1. Where is Earth's water and in what forms does it exist?
2. **Key Concept** Can you think of any other place on Earth where you might find water?

What is freshwater?

Satellite images of Earth show more water than dry land. Most of the water that covers Earth is salt water. Only about 3 percent is **freshwater**—*water that has less than 0.2 percent salt dissolved in it.* Life, as we know it, cannot continue without freshwater.

Water cycles on Earth. Water moves from Earth's surface into the atmosphere by evaporation. The water then condenses and falls back to the surface as precipitation—rain, snow, sleet, or hail. Only freshwater enters Earth's atmosphere and returns to Earth's surface.

More than two-thirds of Earth's freshwater is frozen, as illustrated in **Figure 1.** The rest is liquid water, and most is stored underground. Less than 1 percent of Earth's liquid freshwater is in streams and lakes.

Reading Check Where is Earth's freshwater?

Freshwater on Earth

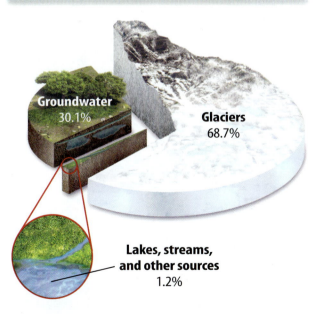

- Groundwater 30.1%
- Glaciers 68.7%
- Lakes, streams, and other sources 1.2%

Figure 1 Most of Earth's freshwater is frozen in glaciers.

Lesson 1
EXPLORE

Glaciers and Ice Sheets

Glaciers are large masses of moving ice that form on land. Glaciers cover about 10 percent of Earth's surface. They are near the North Pole and the South Pole and on mountaintops, as shown in **Figure 2**.

How do glaciers form? Imagine what happens when snow falls but doesn't melt. Year after year layers of snow pile up. The weight and pressure of the snow above compresses the snow on the bottom into ice. Over time, the mass of ice and snow gets so heavy that gravity starts to slowly drag it downhill. For most glaciers this process takes over one hundred years.

Alpine Glaciers

A glacier that forms in the mountains is an **alpine glacier.** Alpine glaciers are on every continent except Australia. They flow downhill like slow-moving rivers of ice. As an alpine glacier flows downhill, it eventually reaches an elevation where temperatures are warm enough to melt the ice. The melted ice is called glacial meltwater.

> **REVIEW VOCABULARY**
> **glacier**
> a large, slow-moving mass of ice and snow

> **WORD ORIGIN**
> **alpine**
> from French *Alpes*, means "Alps"—mountain system of Europe

Figure 2 More than 97 percent of Earth's glacial ice is stored in ice sheets that cover Antarctica and Greenland. Less than 3 percent is stored in alpine glaciers.

Ice Sheets

A glacier that spreads over land in all directions is called an **ice sheet**. Ice sheets are also called continental glaciers. They cover large areas of land (more than 50,000 km^2) and store enormous amounts of freshwater. The only two ice sheets currently on Earth are in Antarctica and Greenland.

Parts of the Antarctic and Greenland ice sheets extend into the ocean. When a glacier flows into the ocean, an ice shelf forms. Ice shelves occur along the coastlines of Alaska, Canada, Greenland, and Antarctica. Icebergs are blocks of ice that break away from ice shelves and float in the ocean.

Antarctic Ice Sheet Earth's largest ice sheet is the Antarctic Ice Sheet. The ice sheet covers most of Antarctica and is larger in surface area than the continental United States. Scientists subdivide the ice sheet into two areas—the West and East Antarctic Ice Sheets—as illustrated in **Figure 3.** The average thickness of the Antarctic Ice Sheet is about 2.4 km. In some places, the ice can be as much as 5 km, or 3 miles, thick.

Greenland Ice Sheet Earth's second-largest ice sheet covers most of Greenland. Its average thickness is about 2.3 km. The total area of the ice sheet is about 1.8 million km^2.

 Reading Check What are the two types of glaciers?

Antarctic Ice Sheet

Math Skills

Volume
How much freshwater is stored as ice in the Antarctic Ice Sheet?

The area (*A*) of the Antarctic Ice Sheet is **14 million km^2**. Calculate the volume (*V*) of ice by multiplying the area by the thickness or height (*h*).

$V = A \times h$

For example, the average thickness of the Antarctic Ice Sheet is **2.4 km**.

V = **14,000,000 km^2** × **2.4 km,** or 33,600,000 km^3.

Practice
The total area of the entire United States is approximately 10 million km^2. What would be the volume of an ice sheet covering the United States to a depth of 2.2 km?

- Review
- Math Practice
- Personal Tutor

Figure 3 Antarctica has an area of about 14 million km^2. That's much larger than the area of the United States, about 10 million km^2. Ice shelves extend into the ocean from several places along the Antarctic coast.

FOLDABLES
Use a sheet of notebook paper to make a two-tab book. Use it to organize your notes about Earth's major forms of frozen water.

How much freshwater is in glaciers?

Glaciers can stay frozen for thousands of years. During periods of Earth's history, the climate was colder than it is now. During those periods, many glaciers formed. The coldest periods are called ice ages—long periods of time when large areas of land are covered by glaciers. The last ice age ended about 10,000 years ago.

Past Changes in Sea Level Even if you have never been to either coast, you probably know that sea level is the average level of the surface of Earth's oceans. Changes in sea level have occurred throughout Earth's history. Sea level rises or falls as climate changes cause the melting or forming of glaciers.

As illustrated in the first image in **Figure 4,** sea level during the last ice age was much lower than it is today. That is because of the enormous amount of Earth's water frozen in vast ice sheets. When the ice sheets melted at the end of the ice age, the meltwater flowed into the ocean and raised sea level.

 Key Concept Check How do glaciers affect sea level?

Melting Glaciers Scientists estimate that if all the glaciers on Earth melted, sea level would rise about 70 meters. Some low-lying areas, such as the Florida peninsula and a large portion of Louisiana, would be under water.

How much water is frozen in the Antarctic ice sheets? The middle image in **Figure 4** illustrates how sea level around the Florida peninsula could change if the West Antarctic ice sheet melted. The last image in **Figure 4** illustrates how sea level for Florida would change if the East Antarctic ice sheet melted.

Changing Sea Level

Figure 4 These maps show the outline of Florida's coast today. The green area in the first illustration shows how much land was above sea level during the last ice age.

20,000 years ago at the height of the last ice age, sea level was about 120 meters lower than it is today.

If the West Antarctic ice sheet melted, sea level would rise about 5 meters above current sea level. The southern tip of Florida would be under water.

If the larger East Antarctic ice sheet melted, sea level could rise by about 51.8 meters. This would put most of Florida under water.

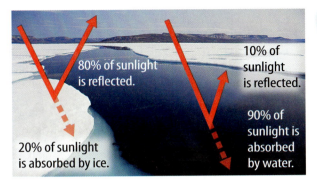

Figure 5 Up to 80 percent of the sunlight that strikes snow or sea ice is reflected into space.

Sea Ice and Snow Cover

Snow and sea ice are also frozen forms of freshwater. **Sea ice** *is ice that forms when seawater freezes.* As seawater freezes, salt is left behind in the ocean. Much of the Arctic Ocean is covered with sea ice.

Unlike glaciers, sea ice does not raise sea level by adding water to the ocean. Consider an ice cube floating in a glass. The amount of water frozen in the ice cube is equal to the amount of water that it displaces in the glass. When the ice cube melts, the water level in the glass stays the same. Likewise, when sea ice melts, sea level stays the same.

However, melting snow or sea ice can affect climate. Snow or ice reflects more solar energy than land or water does. As illustrated in **Figure 5,** most of the sunlight that hits snow or ice is reflected back into space. Reflection helps keep surface temperatures and air temperatures low.

Scientists have recorded a decreasing trend in the amount of snow cover. When snow melts, Earth's surface absorbs more solar energy and heats the air above it. When large areas of Earth's surface are affected over long periods of time, climate changes. Scientists hypothesize that this decrease in snow cover is related to an increase in global temperature.

 Key Concept Check How can sea ice or snow cover affect climate?

Inquiry MiniLab   10 minutes

Does the ground's color affect temperature?

Have you ever been outside on a sunny day after a snowstorm? Did you have to squint because of the glare reflected from the snow? After the snow melts, you do not have to squint as much when you go outside. Why?

1. Read and complete a lab safety form.
2. Lay a sheet of **black paper** and a sheet of **white paper** next to each other on the lab table.

3. Lay one **thermometer** on each sheet of paper. Be sure to place the bulbs of the thermometers on top of the paper.
4. Create a data table to record the temperature of each thermometer, once per minute for 5 minutes.
5. Record the temperature for each sheet of paper.
6. Position a **desk lamp** 20 cm above the thermometers. The lamp should be equidistant from each thermometer bulb. Turn on the lamp.
7. Record the temperature of each thermometer every minute for 5 minutes.

Analyze and Conclude

1. **Graph** your temperature data. Use two colored pencils to differentiate your results. Label each axis and give your graph a title.
2. **Explain** why the temperature readings differed.
3. **Key Concept** In the past, a continental ice sheet covered much of North America. Hypothesize how the melting of the ice sheet affected the temperature of North America.

Figure 6 This graph shows changes in global temperature and atmospheric CO_2 over the past 400,000 years. The steepest rise in CO_2 levels, shown by the red line, began about 150 years ago, when people first began burning fossil fuels. ▶

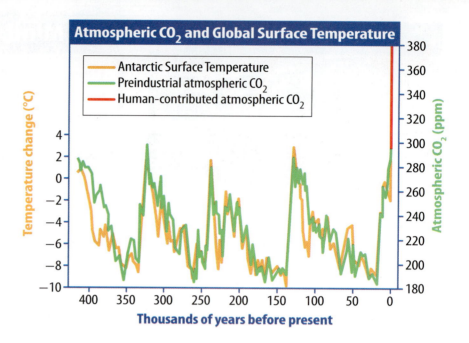

Figure 7 Bubbles of gas locked in Antarctic ice cores provide evidence of the CO_2 content of the atmosphere during periods of Earth's history. ▼

Human Impacts on Glaciers

Scientific studies indicate that Earth's glaciers are melting. Sea ice that covers the Arctic Ocean is also melting. Why? Earth is getting warmer. Data collected by scientists who study Earth's climate show that Earth's average surface temperature has risen approximately 0.5°C since the start of the twentieth century.

Evidence of Climate Change

The orange line in the graph in **Figure 6** represents Earth's average surface temperature during the past 400,000 years. Notice that Earth's temperature fluctuated during that span of time. During cold periods, glaciers formed and sea level fell. During warm periods, glaciers melted and sea level rose.

The green line in **Figure 6** represents the amount of carbon dioxide (CO_2) in Earth's atmosphere. Notice the comparison between Earth's temperature and the amount of atmospheric CO_2. As the amount CO_2 rose, so did Earth's temperature. The data represented by this graph came from **ice cores**—*long columns of ice taken from glaciers* like the one shown in **Figure 7**.

Look again at **Figure 6** and notice the sharp rise in CO_2 shown by the red line. Human activities—especially the burning of fossil fuels—add CO_2 to the atmosphere. Atmospheric CO_2 has risen sharply since the 1800s. Scientists hypothesize that this rise in CO_2 has contributed to the recent rise in global temperature. Many scientists also hypothesize that this rise in temperature is causing many of Earth's glaciers to melt.

Key Concept Check How do human activities affect glaciers?

Melting Glaciers

As Earth's average surface temperature increases, glaciers and ice sheets melt. More water flows into the oceans and sea level rises. **Figure 8** shows how much melting took place in one alpine glacier over a period of 63 years. Like melting ice sheets, melting alpine glaciers contribute to the rise in sea level.

Melting Sea Ice

The Arctic Ocean has been covered with sea ice since the beginning of the last ice age, 125,000 years ago. However, arctic sea ice is melting. **Figure 9** illustrates how much sea ice melted in the Arctic Ocean between 1979 and 2005. It also shows how much arctic sea ice could be lost over the next few decades. In September 2007, sea ice at the North Pole was surrounded by ice-free water for the first time in known human history. Can the melting of snow or ice cause more sea ice to melt?

Positive Feedback Loop

A back-and-forth relationship occurs between melting snow and rising temperature—an increase in one causes an increase in the other. Scientists call this a positive feedback loop. For example, as snow or ice melts, the amount of energy absorbed from the Sun increases. As the amount of energy absorbed from the Sun increases, global temperature rises. As global temperature rises, more snow or ice melts. This repeating cycle is called a positive feedback loop—an increase in one variable causes a corresponding increase in another variable.

Figure 8 Much of the Muir glacier in Alaska has melted since 1941.

Visual Check What changes are visible in the 2004 photo?

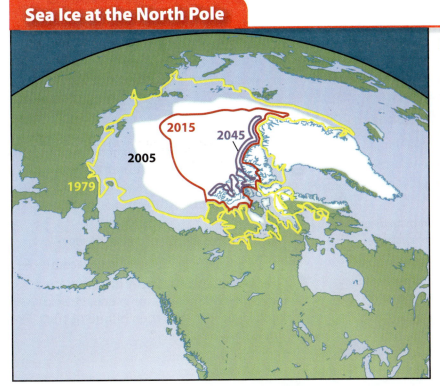

Sea Ice at the North Pole

Figure 9 The white portion of this computer-generated image shows the size of the Arctic Ocean ice cap in 2005. The yellow outline shows the edges of the ice cap in 1979. Red and purple outlines show where scientists predict the edges of the ice cap will be in future years.

Lesson 1 Review

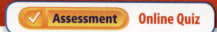 Assessment   Online Quiz

Visual Summary

Most of Earth's freshwater is frozen in ice sheets and alpine glaciers.

Glacier formation can cause sea level to fall. Melting of glaciers can cause sea level to rise.

Human activities are associated with the recent melting of glacial ice and Arctic sea ice.

FOLDABLES

Use your lesson Foldable to review the lesson. Save your Foldable for the project at the end of the chapter.

What do you think NOW?

You first read the statements below at the beginning of the chapter.

1. On Earth, freshwater occurs only as liquid water.
2. Up to 80 percent of the sunlight that strikes snow or ice is reflected back into space.

Did you change your mind about whether you agree or disagree with the statements? Rewrite any false statements to make them true.

Use Vocabulary

1. **Distinguish** between the terms *ice sheet* and *alpine glacier*.

2. **Use the term** *freshwater* in a complete sentence.

3. **Define** *glacier* in your own words.

Understand Key Concepts

4. **Compare and contrast** ice in Antarctica and ice in the Arctic Ocean.

5. Where is most of the freshwater on Earth?
   A. glaciers   C. lakes
   B. groundwater   D. oceans

6. **Contrast** the sea level during the last ice age with today's sea level.

Interpret Graphics

7. **Analyze** The graph below illustrates how the CO_2 content of Earth's atmosphere has changed since 1850. How do these data relate to changes in Earth's temperature?

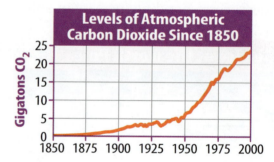

Critical Thinking

8. **Support** Use facts to support this statement: "Freshwater is not evenly distributed on Earth."

Math Skills

Math Practice

9. The Greenland Ice Sheet has an area of 1.8 million km² and an average thickness of 2.3 km. What is the volume of the Greenland Ice Sheet?

614 Chapter 17
EVALUATE

SCIENCE & SOCIETY

Life at the Top of the World

Average temperatures on Earth are increasing; sea ice is melting and disrupting entire ecosystems, which could threaten the survival of polar bears.

Life isn't easy when you're living at the top of the world in the vast, icy region known as the Arctic. It's so cold that ice covers parts of the Arctic Ocean all year long. However, a variety of species thrive in this polar climate. In fact, many ecosystems depend on the Arctic's ice for survival.

But as Earth's average temperatures increase, ice in the Arctic is melting. This includes sea ice—the ice that forms in an ocean. Sea ice follows a natural cycle in the Arctic. It spreads across the Arctic Ocean in winter then decreases in area during summer. But with rising temperatures, sea ice forms later and melts earlier each year. Over the last few decades, the amount of ice in the Arctic has decreased dramatically.

The disappearing ice is threatening the Arctic's top predator, the polar bear. Polar bears travel across sea ice to hunt for seals. As sea ice breaks up and melts, polar bears must swim longer distances to find prey. Also, late freezes and early thaws of sea ice mean shorter hunting seasons for them. Polar bears have been classified as a threatened species because their numbers are decreasing. If warming continues, they could become extinct.

The future of Arctic life is uncertain. Scientists continue to monitor climate data to understand the impact of increasing average temperatures on Arctic ecosystems. But, if Earth's climate continues to warm, life in the Arctic might never be the same.

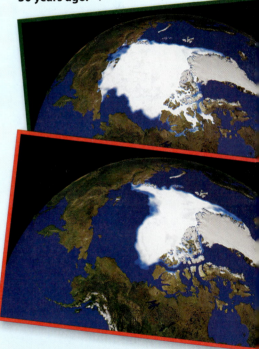

Scientists use satellite images to monitor the amount of Arctic sea ice. These 1979 (top) and 2007 (bottom) images show that the area covered by summer sea ice is about half of what it was over 30 years ago. ▼

◀ Polar bears are strong swimmers and hunt from the ice out at sea as well as from land. During the winter, they build up a layer of fat that helps them survive the rest of the year.

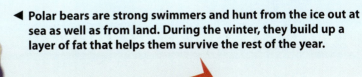

It's Your Turn

RESEARCH Learn how the dwindling population of polar bears would affect other Arctic species. Create a cause and effect diagram with an if-then statement describing the effects of polar bear population decline on other wildlife in the Arctic.

Lesson 2

Streams and Lakes

Reading Guide

Key Concepts 🔑
ESSENTIAL QUESTIONS
- What are streams and lakes?
- What is a watershed?
- How do human activities affect streams and lakes?

Vocabulary
runoff p. 617
stream p. 618
watershed p. 619
estuary p. 619
lake p. 620

🅖 Multilingual eGlossary

▶ Video

What's Science Got to do With It?

Inquiry **What is this structure?**

The large concrete structure shown here is the Hoover Dam, in Nevada. The dam was built to control water flow along the Colorado River. Notice the large reservoir, Lake Mead, behind the dam. Freshwater from Lake Mead is used for recreational purposes, drinking water, irrigation, and hydroelectric power. Dams can also have negative effects on the environment and the ecosystem around a river.

Launch Lab

10 minutes

How can you measure the health of a stream?

The quality of the water in a stream affects the organisms that live in the stream. Macro-invertebrates are tiny animals without backbones. Their presence can be used to determine the health of a stream. For example, the riffle beetle is only in streams where dissolved oxygen is high and the stream is healthy. Use the data below to measure the health of a stream near a new housing development.

1. Read and complete a lab safety form.
2. Use **graph paper** and **colored pencils** to construct a graph using the data provided.
3. Plot the water temperature, dissolved oxygen, and the population density for each year represented.

Year	Water Temp (°C)	Dissolved Oxygen Concentration (ppm)	Riffle Beetle (adults/rock)
1998	10.4	11.5	9.8
2000	11	10.5	9.3
2002	12.7	8	7.9
2004	13.3	7.5	6.2
2006	14.1	6.5	4.4
2008	15.2	5.5	2.6

Think About This

1. What is happening to the stream?
2. Make a prediction about the number of adult riffle beetles per rock in 2015.
3. 🔑 **Key Concept** Other than a decrease in oxygen, what else might affect the riffle beetle population in this stream?

Runoff

If you've ever been outside during a heavy rain, you might have noticed sheets of water rushing downhill over pavement or soil. Water can follow many different paths during a rainstorm. Some water soaks into soil. Some water collects in puddles that evaporate.

Water that flows over Earth's surface is called **runoff**. It comes from rain, melting snow or ice, or any water that does not soak into the soil or evaporate. Runoff is part of the water cycle. Gravity causes runoff to flow downhill, from higher ground to lower ground. Runoff usually starts as a thin layer, or a sheet, of water flowing over the ground, such as the runoff shown in **Figure 10.**

✓ **Reading Check** What is runoff?

Figure 10 Runoff often starts as sheets of water that flow downhill.

▲ Figure 11
Streams form when runoff erodes channels that carry water and sediment downhill.

Streams

A body of water that flows within a channel is a **stream**, as shown in **Figure 11**. Scientists use the term *stream* to refer to any naturally flowing channel of water. For example, a river is a large stream. A brook is a small stream. A creek is larger than a brook but smaller than a river.

 Key Concept Check What is a stream?

All streams form from similar processes. As water flows downhill, it wears away rock and soil, forming tiny channels called rills. Every time it rains, more rock and soil is removed from a rill. Eventually a rill grows in size and forms a larger and more permanent stream channel. Small streams can combine and form a larger stream. Large streams can eventually become rivers that flow into a lake or an ocean.

Pools and Riffles

If you've ever watched a small stream, you might have noticed differences in the way the water flows. Sometimes the water appears smooth, and sometimes it's turbulent or rough, as shown in **Figure 12**. The water is slow, steady, and smooth in places where the stream channel is flat. Pools often form in depressions or low spots within a stream channel. Where the stream channel is rough or the slope is steep, the water tumbles and splashes. A riffle is a shallow part of a stream that flows over uneven ground. Riffles help mix water as it splashes and swirls over rough areas. This action increases the oxygen content of the water and makes the stream healthier.

Figure 12 Oxygen from the air mixes into water as it passes over riffles. Water from riffles helps supply oxygen to the pools downstream. ▼

Watershed

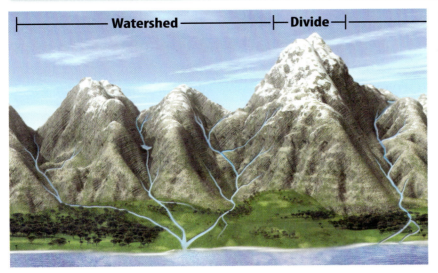

Figure 13 This watershed includes several streams. They flow into a river, which flows into the ocean.

Watersheds

Imagine a house with a roof that is higher in the middle than at its edges. Rain falls on both sides of the roof and runs downward. However, rain runs down the roof in opposite directions. The same thing happens when rain falls to Earth. The direction in which runoff flows depends on which side of a slope the rain falls. A **watershed** *is an area of land that drains runoff into a particular stream, lake, ocean, or other body of water.*

Like the example described above, the boundaries of a watershed are the highest points of land that surround it. These high points are called divides. **Figure 13** shows examples of watersheds and divides.

 Key Concept Check What is a watershed?

From Headwaters to Estuaries

Small streams that form near divides are called headwaters. Streams begin at the headwaters. Streams end at the mouth of a river, where runoff drains into a lake, an ocean, or another large body of water.

What happens when a river meets the sea? Freshwater mixes with salt water. Many large watersheds end in an **estuary**—*a coastal area where freshwater from rivers and streams mixes with salt water from seas or oceans.* Estuaries contain brackish water—a mixture of freshwater and salt water. As **Figure 14** shows, the water in an estuary gets saltier as it gets closer to the ocean. Estuaries are rich in minerals and nutrients and provide important habitats for many organisms.

WORD ORIGIN
estuary
from Latin *aestuarium*, means "a tidal marsh"

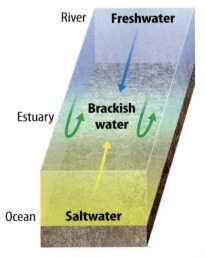

Figure 14 Estuaries form in places where freshwater streams flow into an ocean, a sea, or a bay.

FOLDABLES

Use a sheet of notebook paper to make a two-tab book. Label it as shown. Use it to organize information on the characteristics of streams and lakes.

SCIENCE USE v. COMMON USE

basin
Science Use a shallow depression surrounded by higher ground

Common Use a tub or container used to hold liquids such as water

Lakes

When runoff flows into a **basin**, or a depression in the landscape, a lake can form. *A* **lake** *is a large body of water that forms in a basin surrounded by land.* Most of Earth's lakes are in the Northern Hemisphere. Over 60 percent are in Canada. Lakes are reservoirs that store water. Most lakes contain freshwater.

How Lakes Form

Erosion, landslides, movements of Earth's crust, or the collapse of volcanic cones can form lake basins. Water can enter a lake basin from precipitation, streams, or groundwater that rises to the surface. Most lakes have one or more streams that remove water when the lake overflows. Lakes also lose water by evaporation or when lake water soaks into the ground.

The water level in a lake is not constant. If the lake loses water to evaporation, the lake level will drop. Occasionally a lake will disappear entirely if precipitation does not replenish water lost from the lake. In contrast, if the lake receives too much rainfall, the water can spill over the lake banks and cause a flood.

Inquiry MiniLab

15 minutes

How does a thermocline affect pollution in a lake?

When the Sun warms the surface of a cold lake, a warm layer of water forms on the surface. The denser, cold, deep water remains unchanged. Water of different densities does not mix easily. The pollutants or nutrients in one layer are trapped.

1. Read and complete a lab safety form.
2. Use a **pencil** to poke holes in the bottom of a **paper cup.** Attach the cup to the corner of a **clear plastic shoe box** with **tape.**
3. Fill the shoe box with very **warm water** until the bottom of the cup is submerged.
4. "Pollute" your lake by pouring 100–200 mL of **ice cold water** tinted with **food coloring** into the cup. Observe until water stops flowing. Sketch in your Science Journal what happens.
5. Simulate a storm by blowing across the water's surface.

Analyze and Conclude

1. **Explain** what happened to the tinted water.
2. **Describe** the effect the wind had on the tinted layer.
3. **Contrast** the way the layers formed in your model lake with the way layers form in a real lake.
4. 🔑 **Key Concept** Hypothesize how pollution caused by human activity might affect different areas in a lake.

Properties and Structure

Water changes temperature more slowly than land changes temperature. This can affect weather conditions near the lake. For example, on a hot summer day you might be refreshed by a cool breeze blowing across a lake.

Have you ever been swimming in a lake and noticed that the water changes temperature with depth? Sunlight heats the surface layer, making it warmer and less dense than the layers below. Less sunlight is absorbed the deeper you swim. Some deep, northern lakes develop two **distinct** layers of water—a warm, top layer and a cold, bottom layer. The two layers are separated by a region of rapid temperature change called the thermocline. It acts as a barrier and prevents mixing between the layers.

ACADEMIC VOCABULARY
distinct
(adjective) different or not the same

Human Impact on Streams and Lakes

People worldwide depend on streams and lakes for their water supplies. Streams are dammed to create reservoirs that store water. Because of dams, some rivers, such as the Colorado River in the lesson opener, are nearly dry before they reach the ocean.

As illustrated in **Figure 16,** people can affect the health of streams and lakes in many other ways. Runoff can carry fertilizers, pesticides, sewage, and other pollutants that are harmful to organisms living in or near the water. For example, excess nutrients from fertilizers or sewage can enter a stream and result in an increase in the population of algae. When the algae die, bacteria break down the algae and use oxygen in the decay process. If decay rates are too high, oxygen levels in the water can be so low that fish and other animals cannot survive.

 Key Concept Check How do human activities affect streams and lakes?

Figure 16 Pollutants that flow into freshwater can harm living organisms, including the people who use the water for drinking, washing, and irrigation.

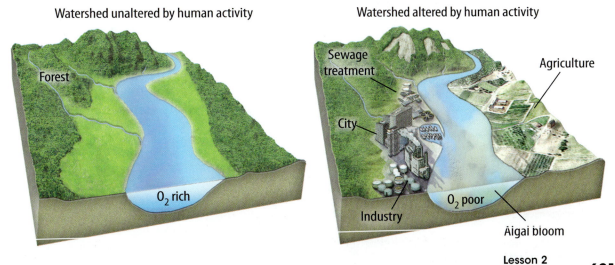

Lesson 2 Review

Visual Summary

Water that flows over Earth's surface and into streams and lakes is called runoff.

Watersheds begin at high places called divides, where headwaters flow downhill.

Humans can have a negative impact on the health of streams and lakes.

FOLDABLES
Use your lesson Foldable to review the lesson. Save your Foldable for the project at the end of the chapter.

What do you think NOW?

You first read the statements below at the beginning of the chapter.

3. Organisms living in a lake or a stream are not affected by the amount of oxygen and nutrients in the water.
4. Glaciers can form lake basins.

Did you change your mind about whether you agree or disagree with the statements? Rewrite any false statements to make them true.

Use Vocabulary

1. **Use the term** *runoff* in a complete sentence.
2. **Define** *watershed* in your own words.
3. **Distinguish** between a *lake* and a *stream*.

Understand Key Concepts

4. Which is the correct order of small bodies of water to large bodies of water?
   A. creeks, rivers, runoff, estuaries
   B. estuaries, runoff, creeks, rivers
   C. rivers, estuaries, runoff, creeks
   D. runoff, creeks, rivers, estuaries

5. **Describe** how lakes form.

6. **Distinguish** between runoff and streams.

7. **Explain** how divides affect water flow.

Interpret Graphics

8. **Summarize** Look at the diagram below. How is this structure formed?

Critical Thinking

9. **Synthesize** Swimmers dip their toes into lake water, and it feels warm. When they dive in, they discover that the lake is much colder. Explain why.

10. **Evaluate** Write a paragraph describing how the destruction of a forest to make room for a factory could affect organisms in a nearby stream.

Inquiry Skill Practice: Observe

30 minutes

How does water flow into and out of streams?

You have probably seen water flow along a stream or a river. What you can see is only part of the story about how water flows through a river, because a lot of the river's flow is in the ground.

Materials

stream table

dry sand

plastic gallon jug

plastic tub

paper towels

Safety

Learn It

Observation is a basic science skill. Without observations, scientists would not know what questions to ask or how to approach and develop ideas about how nature behaves.

Try It

1. Read and complete a lab safety form.

2. Half-fill a stream table with sand. Tilt the table and put the lower end of the drain tube in a plastic tub to allow for drainage.

3. Shape the sand into two long mountain ranges with a valley between them. Reshape the mountains as needed.

4. Poke pin holes in the bottom of a plastic gallon jug and fill it with water. Begin to "rain" on the sand at the top of the stream table. Measure the time it takes for the water to start flowing into the plastic tub as you provide constant rain. Continue until all the sand is wet and the water is flowing steadily.

5. Once all the sand is wet, stop the rain and time how long it takes for the water to stop flowing.

Apply It

6. When you first started the rain on the stream table, where did all of the water go?

7. What had to happen for the water to begin to flow?

8. Once you stopped the rain, why did the water keep draining from the stream table?

9. 🔑 **Key Concept** Where does the water that flows in a stream come from?

Lesson 2
EXTEND
623

Lesson 3

Reading Guide

Key Concepts
ESSENTIAL QUESTIONS
- What is groundwater?
- Why are wetlands important?
- How do human activities affect groundwater and wetlands?

Vocabulary
groundwater p. 625
water table p. 626
porosity p. 626
permeability p. 626
aquifer p. 627
wetland p. 628

 Multilingual eGlossary

 Video BrainPOP®

Groundwater and Wetlands

Inquiry Where did this water come from?

Why is this water bubbling up out of the ground? Have you ever seen anything like it? Groundwater, which is stored in rocks below the surface, can flood the landscape after a severe storm when the ground is saturated. It can also surface in low-lying areas.

Inquiry Launch Lab

10 minutes

How solid is Earth's surface?

It feels solid, but just how solid is "solid ground"? Believe it or not, the soil and rock beneath your feet are not entirely solid.

1. Read and complete a lab safety form.
2. Fill a large, empty **jar** with **golf balls.**
3. Follow the instructions given by your teacher.

Think About This

1. At what point did you think the jar was full?
2. Do you think the particle size of the soil affects how quickly water can move through it?
3. **Key Concept** Does water flow more easily through sediments of equal size or sediments with a variety of different sizes?

Groundwater

Some water that falls to Earth as precipitation soaks into the ground. *Generally, water that lies below ground is called* **groundwater.** Water seeps through soil and into tiny pores, or spaces, between sediment and rock. If you have ever been inside a cave and seen water dripping down the sides, you've seen groundwater seeping through rock.

Key Concept Check What is groundwater?

In some areas, groundwater is very close to the surface and keeps the soil wet. In other areas, especially deserts and other dry climates, groundwater is hundreds of meters below the surface.

Groundwater can remain underground for long periods of time—thousands or millions of years. Eventually, it returns to the surface and reenters the water cycle. Humans interfere with this process, however, when they drill wells into the ground to remove water for everyday use.

Importance of Groundwater

The water beneath Earth's surface is much more plentiful than the freshwater in lakes and streams. Recall that groundwater is about one-third of Earth's freshwater. Groundwater is an important source of water for many streams, lakes, and wetlands. Some plant species absorb groundwater through long roots that grow deep underground.

People in many areas of the world rely on groundwater for their water supply. In the United States, about 20 percent of the water people use daily comes from groundwater.

Groundwater

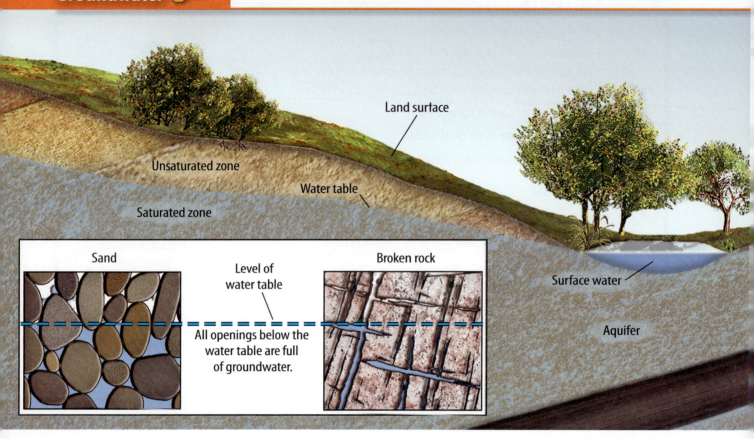

The Water Table

As illustrated in **Figure 17,** groundwater seeps into tiny cracks and pores within rocks and sediment. Near Earth's surface, the pores contain a mixture of air and water. This region is called the unsaturated zone. It is called unsaturated because the pores are not completely filled with water. Farther beneath the surface, the pores are completely filled with water. This region is called the saturated zone. *The upper limit of the saturated zone is called the* **water table.**

Reading Check What is the water table?

Porosity Rocks vary in the amount of water they can hold and the speed with which water flows through the rock. Some rocks can hold a lot of water and some rocks cannot. **Porosity** *is a measure of rock's ability to hold water.* Porosity increases with the number of pores in the rock. The higher the porosity, the more water a rock can contain.

Permeability *The measure of water's ability to flow through rock and sediment is called* **permeability.** This ability to flow through rock and sediment depends on pore size and the connections between the pores. Even if pore space is abundant in a rock, the pores must form connected pathways for water to flow easily through the rock.

Groundwater Flow

Just as runoff flows downhill across Earth's surface, groundwater flows downhill beneath Earth's surface. Groundwater flows from higher elevations to lower elevations. In low-lying areas at Earth's surface, groundwater might eventually seep out of the ground and into a stream, a lake, or a wetland, as also shown in **Figure 17.** In this way, groundwater can become surface water. Likewise, surface water can seep into the ground and become groundwater. This is how groundwater is replenished.

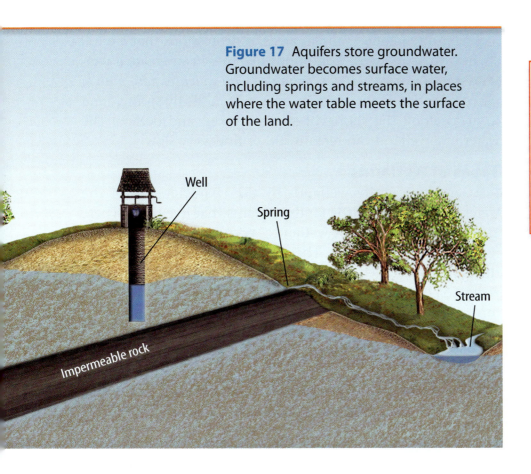

Figure 17 Aquifers store groundwater. Groundwater becomes surface water, including springs and streams, in places where the water table meets the surface of the land.

FOLDABLES

Use a sheet of notebook paper to make a two-tab book. Use it to organize your notes on groundwater, wetlands, and how each relates to the other.

Wells People often bring groundwater to Earth's surface by drilling wells like the one shown in **Figure 17**. Wells are usually drilled into an **aquifer**—*an area of permeable sediment or rock that holds significant amounts of water*. Groundwater then flows into the well from the aquifer and is pumped to the surface.

Precipitation helps replace groundwater drawn out of wells. During a drought, less groundwater is replaced, so the water level in a well drops. The same thing happens if water is removed from a well faster than it is replaced. If the water level drops too low, a well runs dry.

Springs A spring forms where the water table rises to Earth's surface, as shown in the lesson opener. Some springs bubble to Earth's surface only after heavy rain or snowmelt. Many springs fed by large aquifers flow continuously.

Human Impact on Groundwater

If polluted surface water seeps into the ground, it can pollute the groundwater below it. Pollutants include pesticides, fertilizers, sewage, industrial waste, and salt used to melt ice on highways. Pollutants can travel through the ground and into aquifers that supply wells. People's health can be harmed if they drink contaminated water from a well.

The water in an aquifer helps to support the rocks and soil above it. In some parts of the world, water is being removed from aquifers faster than it can be replaced. This creates empty space underground. The empty space underground cannot support the weight of the overlying rock and soil. Sinkholes form where the ground collapses due to lack of sufficient support from below.

 Key Concept Check How do human activities affect groundwater?

Lesson 3
EXPLAIN

Wetlands

Water often collects in flat areas or depressions that are too shallow to form lakes. Conditions like these can create a **wetland**—*an area of land that is saturated with water for part or all of the year.* Wetlands also form in areas kept moist by springs, and in areas along the shores of streams, lakes, and oceans. The water in a wetland can remain still or flow very slowly.

Types of Wetlands

Scientists identify wetlands by the characteristics of the water and soil and by the kinds of plants that live there. There are three major types of wetlands, as shown in **Table 1.** Bogs form in cool, wet climates. They produce a thick layer of peat—the partially decayed remains of **sphagnum** moss. Peat holds water, so bogs rarely dry out. Unlike bogs, marshes and swamps form in warmer, drier climates and do not produce peat. Marshes and swamps are supplied by precipitation and runoff. They can temporarily dry out in hot, dry weather.

WORD ORIGIN
sphagnum
from Greek *sphagnos*, means "a spiny shrub"

Table 1 Types of Wetlands

Bogs
- supplied by runoff, low oxygen content
- soil acidic, nutrient-poor
- dominant plants—*Sphagnum* moss, wildflowers, cranberries

Marshes
- supplied by runoff and precipitation
- soil slightly acidic, nutrient-rich
- dominant plants—grasses and shrubs

Swamps
- supplied by runoff and precipitation
- soil slightly acidic, nutrient-rich
- dominant plants—trees and shrubs

Importance of Wetlands

Wetlands provide important habitat for plants and wildlife. They help control flooding and erosion and also help filter sediments and pollutants from water.

Habitat A wide variety of plants and animals live in wetlands, as shown in **Figure 18.** Wetlands provide plentiful food and shelter for young and newly hatched animals, including fish, amphibians, and birds. Wetlands are also important rest stops and food sources for migrating animals, especially birds.

Flood Control Wetlands help reduce flooding because they store large quantities of water. They fill with water during the wet season and release the water slowly during times of drought.

Erosion Control Coastal wetlands help prevent beach erosion. Wetlands can reduce the energy of wave action and storm surges—water pushed onto the shore by strong winds produced by severe storms.

Filtration Wetlands help keep sediments and pollutants from reaching streams, lakes, groundwater, or the ocean. They are natural filtration systems. Runoff that enters a wetland often contains excess nitrogen from fertilizers or animal waste. Plants and the bacteria in wetland soils absorb excess nitrogen. Wetland plants and soils also trap sediments and help remove toxic metals and other pollutants from the water.

 Key Concept Check Why are wetlands important?

Figure 18 Wetlands are an important habitat for wildlife, providing water, food, and shelter.

Inquiry MiniLab

20 minutes

Can you model freshwater environments?

Sources of groundwater and surface water are similar. When the ground becomes saturated with water following a storm, the excess water can flow into lakes or streams. Groundwater may also surface and create wetlands.

1. Read and complete a lab safety form.
2. Half fill a **plastic shoebox** with **sand**. Flatten out the surface of the sand.
3. While watching the side of the box as you pour, add enough water so the water level, or water table, is even with the top of the sand.
4. Scoop some sand from one end of the box and pile the excess sand on the other end of the box to form highlands. Create an area between the scooped hole and the pile where the level of the sand is at the water table.
5. View the plastic shoebox from the side. Draw a cross section of your model in your Science Journal. Include the water table on your cross section.
6. Add a few drops of **food coloring** to the highlands to represent pollution. Using a **paper cup** with holes in the bottom, model precipitation by dripping water down over your model. Observe.

Analyze and Conclude

1. **Label** the lake, wetlands, highlands and water table on your diagram and identify where the water table is above, at the same level as, and below the surface.
2. **Describe** the path the "pollution" follows.
3. **Key Concept** How does this lab demonstrate the behavior of groundwater?

Human Impact on Wetlands

Many wetlands throughout the world have been drained and filled with soil for roads, buildings, airports, and other uses. The disappearance of wetlands has also been associated with rising sea level, coastal erosion, and the introduction of species that are not naturally found in wetlands. Scientists estimate that more than half of all wetlands in the United States have been destroyed over the past 300 years.

The Louisiana coastline has lost about 310 km^2 of wetlands since 1950. **Figure 19** shows some of the changes that have taken place on the Louisiana coast. A lack of coastal wetlands may have contributed to widespread flooding that occurred when Hurricane Katrina struck the area in 2005.

Key Concept Check How do human activities affect wetlands?

Figure 19 Large portions of the coastal wetlands of Louisiana have been filled in to build roads and buildings. Scientists and engineers have proposed that future flooding could be avoided by restoring some wetland areas.

Chapter 17
EXPLAIN

Lesson 3 Review

Visual Summary

Groundwater fills pores in soil and rocks in the saturated zone. The water table marks the top of the saturated zone.

A wetland is an area of land that is saturated with water for part or all of the year.

Human involvement in alteration and destruction of wetlands can have devastating effects.

FOLDABLES

Use your lesson Foldable to review the lesson. Save your Foldable for the project at the end of the chapter.

What do you think NOW?

You first read the statements below at the beginning of the chapter.

5. People use groundwater as a source of water.
6. Wetlands can naturally filter pollutants from groundwater.

Did you change your mind about whether you agree or disagree with the statements? Rewrite any false statements to make them true.

Use Vocabulary

1. **Distinguish** between *porosity* and *permeability*.

2. **Use the term** *groundwater* in a complete sentence.

3. **Define** *water table* in your own words.

Understand Key Concepts

4. Where is the saturated zone located?
   A. above the water table
   B. below the water table
   C. beside the water table
   D. within the water table

5. **Explain** how groundwater can become surface water.

6. **Explain** how groundwater that is used as drinking water reaches homes.

7. **List** three reasons wetlands are important to people and the environment.

Interpret Graphics

8. **Organize Information** Copy and fill in the graphic organizer below to describe the different types of wetlands.

Wetland Type	Description
Marsh	
Swamp	
Bog	

Critical Thinking

9. **Design an experiment** to test whether groundwater from a well has been contaminated by wastewater from a nearby sewage treatment plant.

10. **Explain** A local developer wants to drain a wetland in your city and build a shopping mall in its place. Write a letter to the editor of your local newspaper to explain the possible outcome of the development.

Inquiry Lab

40 minutes

Materials

stream table

tub

sand

food coloring

pencil, round

paper towels

gallon jug

Also needed: Assorted items for reducing pollution, such as plastic wrap, plastic spoons, straws, gravel, paper towels, cat litter pellets, activated charcoal

Safety

What can be done about pollution?

Freshwater is everywhere—running along Earth's surface in rivers, pooling in lakes, and flowing through rocks underground. When pollution from human activity enters the freshwater supply, it quickly spreads through lakes, rivers, and groundwater.

Question

How does pollution from human activity affect Earth's freshwater?

Procedure

1. Read and complete a lab safety form.
2. Half-fill your stream table with sand. Tilt the table and put the drain tube in the tub. Keep the drain clear during lab.
3. Shape the sand into two long mountain ranges with a valley between them. Reshape mountains as needed.
4. Poke pin holes in the bottom of a plastic gallon jug. Fill the jug with water and provide rain until all the sand is wet and the river is flowing continuously.
5. Select three locations on the side of a mountain. Put 10 drops of food coloring around each site to represent a pollution source.
6. Resume rain and observe how the pollution spreads. Use a pencil to drill test wells around a site. Lower a strip of paper towel into the well and wet it in the groundwater to look for pollution. Record your observations in your Science Journal.
7. Once your initial pollution has washed away, add three more polluted locations. Record how long it takes for the pollution to appear and then for the river to run clear again.

Superfund Site

 **Extension**

Find out what a superfund site is and research local sites that have this designation. What are local authorities doing to help clean up the superfund site?

8. Develop a plan for reducing the effects of the pollution. Present your plan to your teacher.

9. Once your plan is approved, introduce pollution in three locations. Help the pollution sink in with a little rain.

10. Implement your plan to reduce the pollution at all three locations.

11. Resume rain. Observe the spread of the pollution, the time it takes the pollution to reach the river, and the time it takes for the river to run clear.

Lab Tips

- ☑ To help keep drain clear, keep the sand several inches back from the drain.
- ☑ Use different colors of food coloring to tell the difference between different pollution sites.

Analyze and Conclude

12. **The Big Idea** On your stream table, identify some of the different places where Earth's freshwater occurs, including, but not limited to lakes, groundwater, rivers, and wetlands.

13. **Describe** How did the pollution spread through the river system?

14. **Compare and Contrast** How did the pollution control sites compare with the uncontrolled sites? Did your plan work? Why or why not?

Communicate Your Results

Create a graph comparing the duration of the pollution for the run with no pollution control measures to the run with pollution control measures. Present your results.

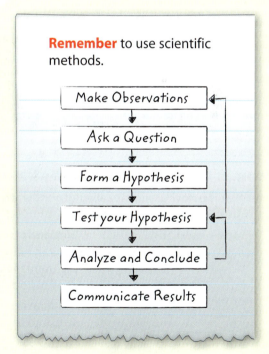

Remember to use scientific methods.

Lesson 3
EXTEND
633

Chapter 17 Study Guide

 WebQuest

 THE BIG IDEA: Earth's freshwater is stored in glaciers, groundwater, lakes, and streams.

Key Concepts Summary

Vocabulary

Lesson 1: Glaciers and Polar Ice Sheets

- **Freshwater** contains less than 0.2 percent salt. More than two-thirds of Earth's freshwater is frozen in ice.
- Snow and ice reflect sunlight and help keep Earth's surface temperatures and air temperatures low.
- Increasing amounts of carbon dioxide in the atmosphere raise temperatures and contribute to the melting of ice and snow.

freshwater p. 607
alpine glacier p. 608
ice sheet p. 609
sea ice p. 611
ice core p. 612

Lesson 2: Streams and Lakes

- A **stream** is a body of water that flows within a channel. A **lake** is a large body of water that forms in a basin or a shallow depression that is surrounded by land. Streams and lakes make up less than 1 percent of Earth's freshwater.
- A **watershed** is an area of land that drains **runoff** into a stream, a lake, an ocean, or another body of water.
- Runoff can carry fertilizers, sewage, pesticides, and other harmful materials into streams and lakes.

runoff p. 617
stream p. 618
watershed p. 619
estuary p. 619
lake p. 620

Lesson 3: Groundwater and Wetlands

- Water that is below ground is called **groundwater.** Almost one-third of Earth's freshwater is groundwater.
- **Wetlands** provide valuable wildlife habitat, help filter sediment and pollutants from runoff, and help control flooding and erosion.
- Pollution from human activities can seep into the ground and contaminate groundwater. Many wetlands have been drained of water and filled with soil to make dry land for roads, buildings, and other uses. Wetlands that are destroyed can no longer filter the runoff that seeps into groundwater.

groundwater p. 625
water table p. 626
porosity p. 626
permeability p. 626
aquifer p. 627
wetland p. 628

634 • Chapter 17 Study Guide

Study Guide

- **Personal Tutor**
- **Vocabulary eGames**
- **Vocabulary eFlashcards**

Chapter Project

Assemble your lesson Foldables as shown to make a Chapter Project. Use the project to review what you have learned in this chapter.

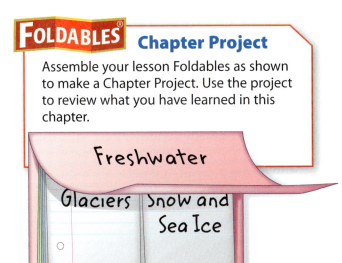

Use Vocabulary

1. Only about 3 percent of Earth's water is _____.

2. Ice formed in the ocean is called _____.

3. A(n) _____ is a large mass of ice that moves slowly over land.

4. Water that flows across Earth's surface is called _____.

5. A(n) _____ is an area of land where all of the water both above and below the ground drains to the same place.

6. A(n) _____ is a body of water surrounded by land and typically contains freshwater.

7. _____ is a measure of a rock's ability to hold water.

8. _____ is the ability of a rock to allow water to pass through it.

Link Vocabulary and Key Concepts Interactive Concept Map

Copy this concept map, and then use vocabulary terms from the previous page to complete the concept map.

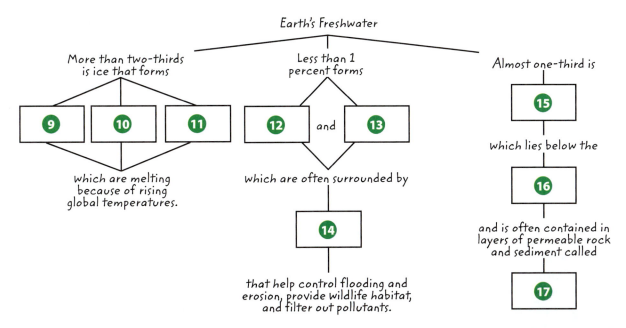

Chapter 17 Review

Understand Key Concepts

1. Alpine glaciers are in
   A. ice sheets.
   B. ice shelves.
   C. mountain valleys.
   D. oceans.

2. What happens when glaciers melt?
   A. More heat energy from the Sun is reflected.
   B. Global temperature decreases.
   C. Sea level rises.
   D. The amount of sea ice increases.

3. Where in the world is the largest area covered by ice sheets?
   A. Antarctica
   B. Asia
   C. Canada
   D. Greenland

Use the following graph to answer the question below.

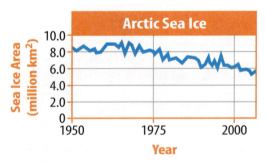

4. How did the area of sea ice in the Arctic change from 1950 to 2000?
   A. approached zero
   B. decreased
   C. increased
   D. stayed the same

5. What causes water to move from areas of higher elevation to areas of lower elevation?
   A. erosion
   B. thermocline
   C. gravity
   D. mixing

6. Which statement about watersheds is true?
   A. All watersheds are the same size.
   B. Larger watersheds can be broken down into smaller watersheds.
   C. Watersheds do not go across state borders.
   D. Watersheds consist only of surface water.

7. Which statement about lakes is NOT true?
   A. Lakes always contain freshwater.
   B. Streams flow in and out.
   C. They form in a hollow basin.
   D. Water temperature varies with depth.

8. Which is NOT a reason wetlands are important?
   A. affect climate
   B. control erosion
   C. filter out pollutants
   D. provide wildlife habitat

9. Which type of wetland is shown in the photo below?
   A. bog
   B. fen
   C. marsh
   D. swamp

10. How does a well run dry?
    A. Groundwater is removed more slowly than it is replaced.
    B. Groundwater is removed more quickly than it is replaced.
    C. Groundwater gets trapped in an aquifer.
    D. Groundwater becomes contaminated with pollutants.

Chapter Review

Assessment
Online Test Practice

Critical Thinking

11 **Explain** how sea ice forms.

12 **Identify** and describe two types of glaciers. Refer to the photos below as examples.

13 **Summarize** how glaciers and sea ice are changing because of human activities.

14 **Effect** How does the melting of alpine glaciers affect humans?

15 **Evaluate** How are scientists using data on changes to snow and ice cover as evidence of global climate change?

16 **Illustrate** the formation of a stream.

17 **Design a model** that shows how a lake can form from the movement of tectonic plates.

18 **Infer** why wetlands are good habitats for newly hatched and juvenile animals, such as fish and amphibians.

19 **Explain** how humans affect wetlands.

20 **Compare and contrast** an aquifer and a spring.

21 **Design a model** that would help you explain the difference between porosity and permeability to a group of elementary school students.

Writing in Science

22 **Write** a five-sentence paragraph describing changes in the amount of sea ice in the Arctic Ocean since 1979. Be sure to include a topic sentence in your paragraph.

REVIEW THE BIG IDEA

23 Where is Earth's freshwater? Describe its general form, such as ice or liquid water. Also describe its general location, such as above ground, below ground, etc. Approximately what percentage of Earth's freshwater is at these locations?

24 What type of wetland appears in the photo below?

Math Skills

Review Math Practice

25 In 2008, a chunk of ice with an area of 570 km^2 broke off an ice shelf in Antarctica. Assuming the ice had an average thickness of 2.4 km, what volume of ice did this ice chunk contain?

26 Water expands when it freezes. If a cubic kilometer of ice produces .91 km^3 of water when it melts, what volume of water would be produced if the ice chunk melted completely?

27 The volume of Lake Erie is 484 km^3. What does this suggest about the effect the melting ice shelf would have on the ocean?

28 There are 63 glaciers in the Wind River Range of Wyoming with a total area of 44.5 km^2. The average thickness of the glaciers is 52.0 meters. What is the average thickness of the glaciers in km? (1 km = 1000 m)

Standardized Test Practice

Record your answers on the answer sheet provided by your teacher or on a sheet of paper.

Multiple Choice

1. Which statement is true of groundwater?
   - A It creates pores between rock and sediment.
   - B It eventually returns to the surface.
   - C It flows quickly uphill in porous soil.
   - D It remains underground very briefly.

Use the diagram below to answer question 2.

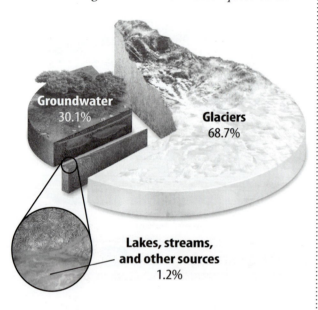

2. What percentage of freshwater comes from glaciers?
   - A 1.2 percent
   - B 30.1 percent
   - C 68.7 percent
   - D 100 percent

3. Which results from crop fertilizer runoff into streams?
   - A algal bloom
   - B crop failure
   - C ozone destruction
   - D weed growth

Use the diagram below to answer question 4.

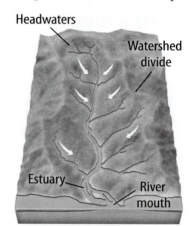

4. Which is correct based on the diagram above?
   - A Watershed divides are at the bases of hills or mountains.
   - B Water in a watershed flows downhill into rivers.
   - C Water in the streams moves from estuary to the headwaters.
   - D Water outside the watershed flows directly into the ocean.

5. How does the reduction of ice and snow on Earth's surface affect climate?
   - A Land and water absorb more solar radiation.
   - B Sea levels decrease because more moisture enters the air.
   - C Air becomes drier because more water runs off the surface.
   - D Global temperatures cool as more of the Sun's heat is absorbed.

6. How do bogs differ from marshes and swamps?
   - A Bogs produce no peat.
   - B Bogs rarely dry out.
   - C Bogs need hot weather.
   - D Bogs are nutrient-rich.

Standardized Test Practice

✓ Assessment
Online Standardized Test Practice

Use the diagram below to answer question 7.

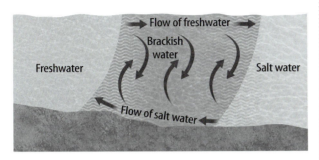

7 What forms from the mixture of freshwater and salt water?

A estuary

B lake

C ocean

D river

8 Which lists naturally flowing channels of water from largest to smallest?

A brook, creek, river

B creek, brook, river

C river, brook, creek

D river, creek, brook

9 Which measures water's ability to flow through rock and sediment?

A fluidity

B liquidity

C permeability

D porosity

Constructed Response

Use the diagram below to answer questions 10 and 11.

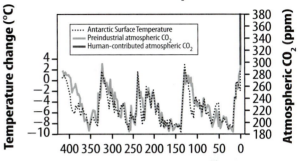

10 Using the graph above, describe the changes in global surface temperatures and CO_2 levels over time. What is a logical inference about climate based on this information?

11 When did Earth's CO_2 levels begin to rise sharply? What human activity contributed to this change? Predict future Antarctic surface temperatures based on this trend.

Use the table below to answer question 12.

Benefit	Explanation
Erosion control	
Filtration	
Flood control	
Habitat	

12 Explain how each benefit listed in the table helps Earth and its inhabitants.

NEED EXTRA HELP?												
If You Missed Question...	1	2	3	4	5	6	7	8	9	10	11	12
Go to Lesson...	3	1	2	2	1	3	2	2	3	1	1	3

Chapter 17 Standardized Test Practice • **639**

Chapter 18

Natural Resources

THE BIG IDEA Why is it important to manage natural resources wisely?

Inquiry What do these colors mean?

This image shows where thermal energy escapes from the inside of a house. Red and yellow areas represent the greatest loss. Blue areas represent low or no loss.

- Which energy resources are used to heat this house?
- Why is it important to reduce thermal energy loss from houses, cars, or electrical appliances?
- Why is it important to manage natural resources wisely?

Get Ready to Read

What do you think?

Before you read, decide if you agree or disagree with each of these statements. As you read this chapter, see if you change your mind about any of the statements.

1. Nonrenewable energy resources include fossil fuels and uranium.
2. Energy use in the United States is lower than in other countries.
3. Renewable energy resources do not pollute the environment.
4. Burning organic material can produce electricity.
5. Cities cover most of the land in the United States.
6. Minerals form over millions of years.
7. Humans need oxygen and water to survive.
8. About 10 percent of Earth's total water can be used by humans.

 Your one-stop online resource

connectED.mcgraw-hill.com

 Video   WebQuest

 Audio   Assessment

 Review   Concepts in Motion

 Inquiry   Multilingual eGlossary

Lesson 1

Reading Guide

Key Concepts 🔑
ESSENTIAL QUESTIONS

- What are the main sources of nonrenewable energy?
- What are the advantages and disadvantages of using nonrenewable energy resources?
- How can individuals help manage nonrenewable resources wisely?

Vocabulary
nonrenewable resource p. 643
renewable resource p. 643
nuclear energy p. 647
reclamation p. 649

🅖 Multilingual eGlossary

Energy Resources

Inquiry What's in the pipeline?

The Trans-Alaska Pipeline System carries oil more than 1,200 km from beneath Prudhoe Bay, Alaska, to the port city of Valdez, Alaska. How might the pipeline's construction and operation affect the habitats and the organisms living along it? How do getting and using fossil fuels impact the environment?

Launch Lab — 20 minutes

How do you use energy resources?
In the United States today, the energy used for most daily activities is easily available at the flip of a switch or the push of a button. How do you use energy in your daily activities?

1. Design a three-column data chart in your Science Journal. Title the columns *Activity*, *Type of Energy Used*, and *Amount of Time*.
2. Record every instance that you use energy during a 24-hr period.
3. Total your usage of the different forms of energy, and record them in your Science Journal.

Think About This
1. How many times did you use each type of energy?
2. Compare and contrast your usage with that of other members of your class.
3. **Key Concept** Are there instances of energy use when you could have conserved energy? Explain how you would do it.

Sources of Energy

Think about all the times you use energy in one day. Are you surprised by how much you depend on energy? You use it for electricity, transportation, and other needs. That is one reason it is important to know where energy comes from and how much is available for humans to use.

 lists different energy sources. Most energy in the United States comes from nonrenewable resources. **Nonrenewable resources** are resources that are used faster than they can be replaced by natural processes. Fossil fuels, such as coal and oil, and uranium, which is used in nuclear reactions, are both nonrenewable energy resources.

Renewable resources are resources that can be replaced by natural processes in a relatively short amount of time. The Sun's energy, also called solar energy, is a renewable energy resource. You will read more about renewable energy resources in Lesson 2.

Key Concept Check What are the main nonrenewable energy resources?

Table 1 Energy resources can be nonrenewable or renewable.

Table 1 Energy Sources	
Nonrenewable Energy Resources	Renewable Energy Resources
fossil fuels uranium	solar wind water geothermal biomass

WORD ORIGIN
resource
from Latin *resurgere*, means "to rise again"

Lesson 1
EXPLORE
643

Nonrenewable Energy Resources

You might turn on a lamp to read, turn on a heater to stay warm, or ride the bus to school. In the United States, the energy to power lamps, heat houses, and run vehicles probably comes from nonrenewable energy resources, such as fossil fuels.

Fossil Fuels

Coal, oil, also called petroleum, and natural gas are fossil fuels. They are nonrenewable because they form over millions of years. The fossil fuels used today formed from the remains of prehistoric organisms. The decayed remains of these organisms were buried by layers of sediment and changed chemically by extreme temperatures and pressure. The type of fossil fuel that formed depended on three factors:

- the type of organic matter
- the temperature and pressure
- the length of time that the organic matter was buried

Reading Check What factors determine which type of fossil fuel forms?

Coal Earth was very different 300 million years ago, when the coal used today began forming. Plants, such as ferns and trees, grew in prehistoric swamps. As shown in **Figure 1,** the first step of coal formation occurred when those plants died.

Bacteria, extreme temperatures, and pressure acted on the plant remains over time. Eventually a brownish material, called peat, formed. Peat can be used as a fuel. However, peat contains moisture and produces a lot of smoke when it burns. As shown in **Figure 1,** peat eventually can change into harder and harder types of coal. The hardest coal, anthracite, contains the most carbon per unit of volume and burns most efficiently.

Figure 1 Much of the coal used today began forming more than 300 million years ago from the remains of prehistoric plants.

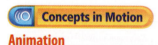
Animation

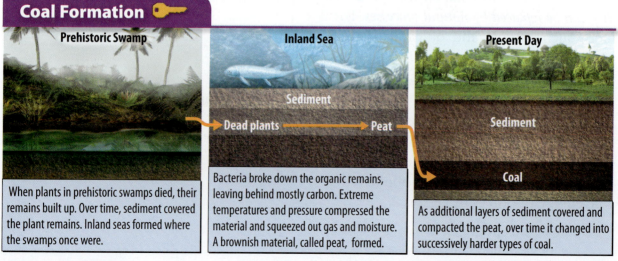

Coal Formation

Prehistoric Swamp
When plants in prehistoric swamps died, their remains built up. Over time, sediment covered the plant remains. Inland seas formed where the swamps once were.

Inland Sea
Bacteria broke down the organic remains, leaving behind mostly carbon. Extreme temperatures and pressure compressed the material and squeezed out gas and moisture. A brownish material, called peat, formed.

Present Day
As additional layers of sediment covered and compacted the peat, over time it changed into successively harder types of coal.

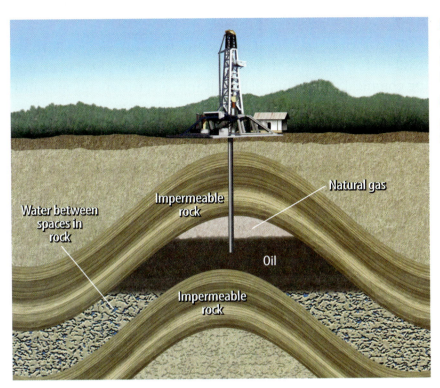

Figure 2 Reservoirs of oil and natural gas often are under layers of impermeable rock.

Visual Check What prevents oil and natural gas from rising to the surface?

Oil and Natural Gas Like coal, the oil and natural gas used today formed millions of years ago. The process that formed oil and natural gas is similar to the process that formed coal. However, oil and natural gas formation involves different types of organisms. Scientists theorize that oil and natural gas formed from the remains of microscopic marine organisms called plankton. The plankton died and fell to the ocean floor. There, layers of sediment buried their remains. Bacteria decomposed the organic matter, and then pressure and extreme temperatures acted on the sediments. During this process, thick, liquid oil formed first. If the temperature and pressure were great enough, natural gas formed.

Most of the oil and natural gas used today formed where forces within Earth folded and tilted thick rock layers. Often hundreds of meters of sediments and rock layers covered oil and natural gas. However, oil and natural gas were less dense than the surrounding sediments and rock. As a result, oil and natural gas began to rise to the surface by passing through the pores, or small holes, in rocks. As shown in **Figure 2,** oil and natural gas eventually reached layers of rock through which they could not pass, or impermeable rock layers. Deposits of oil and natural gas formed under these impermeable rocks. The less-dense natural gas settled on top of the denser oil.

Reading Check How is coal formation different from oil formation?

Lesson 1
EXPLAIN

Advantages of Fossil Fuels

Do you know that fossil fuels store chemical energy? Burning fossil fuels transforms this energy. The steps involved in changing chemical energy in fossil fuels into electric energy are fairly easy and direct. This process is one advantage of using these nonrenewable resources. Also, fossil fuels are relatively inexpensive and easy to transport. Coal is often transported by trains, and oil is transported by pipelines or large ships called tankers.

Disadvantages of Fossil Fuels

Although fossil fuels provide energy, there are disadvantages to using them.

Limited Supply One disadvantage of fossil fuels is that they are nonrenewable. No one knows for sure when supplies will be gone. Scientists estimate that, at current rates of consumption, known reserves of oil will last only another 50 years.

Habitat Disruption In addition to being nonrenewable, the process of obtaining fossil fuels disturbs environments. Coal comes from underground mines or strip mines, such as the one shown in **Figure 3.** Oil and natural gas come from wells drilled into Earth. Mines in particular disturb habitats. Forests might be fragmented, or broken into areas of trees that are no longer connected. Fragmentation can negatively affect birds and other organisms that live in forests.

Reading Check How much longer are known oil reserves predicted to last?

Figure 3 Strip-mining involves removing layers of rock and soil to reach coal deposits.

Pollution Another disadvantage of fossil fuels as an energy resource is pollution. For example, runoff from coal mines can pollute soil and water. Oil spills from tankers can harm living things, such as the bird shown in **Figure 4.**

Pollution also occurs when fossil fuels are used. Burning fossil fuels releases chemicals into the atmosphere. These chemicals react in the presence of sunlight and produce a brownish haze. This haze can cause respiratory problems, particularly in young children. The chemicals also can react with water in the atmosphere and make rain and snow more acidic. The acidic precipitation can change the chemistry of soil and water and harm living things.

Figure 4 One disadvantage of fossil fuels is pollution, which can harm living things. This bird was covered with oil after an oil spill.

Key Concept Check What is one advantage and one disadvantage of using fossil fuels?

Nuclear Energy

Atoms are too small to be seen with the unaided eye. Even though they are small, atoms can release large amounts of energy. *Energy released from atomic reactions is called* **nuclear energy**. Stars release nuclear energy by fusing atoms. The type of nuclear energy used on Earth involves a different process.

Inquiry MiniLab   20 minutes

What is your reaction?

When atoms split during nuclear fission, the chain reaction releases thermal energy and by-products. What happens when your class participates in a simulation of a nuclear reaction?

1. Read and complete a lab safety form.
2. Use a **marker** to label three **sticky notes.** Label one note *U-235.* Label two notes *Neutron.* Stick the U-235 note on your **apron.** Hold the Neutron notes in one hand and a **Thermal Energy Card** in the other. You now represent a uranium-235 atom.
3. When you are tagged with a Neutron label from another student, tag two other student U-235 atoms with your Neutron labels. Drop your Thermal Energy Card into the **Energy Box.**
4. Observe as the remainder of the U-235 atoms are split, and imagine this happening extremely fast at the atomic level.

Analyze and Conclude

1. **Describe** what the simulation illustrated about nuclear fission.
2. **Predict** what would happen if, in the simulation, your classroom was filled wall-to-wall with U-235 atoms and the chain reaction got out of control.
3. **Key Concept** Identify one advantage and one disadvantage of nuclear energy.

Lesson 1

647

EXPLAIN

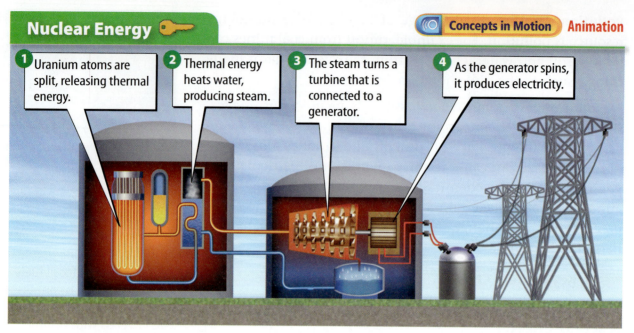

Figure 5 In a nuclear power plant, thermal energy released from splitting uranium atoms is transformed into electrical energy.

Nuclear Fission Nuclear power plants, such as the one shown in **Figure 5,** produce electricity using nuclear fission. This process splits atoms. Uranium atoms are placed into fuel rods. Neutrons are aimed at the rods and hit the uranium atoms. Each atom splits and releases two to three neutrons and thermal energy. The released neutrons hit other atoms, causing a chain reaction of splitting atoms. Countless atoms split and release large amounts of thermal energy. This energy heats water and changes it to steam. The steam turns a turbine connected to a generator, which produces electricity.

 Reading Check What are the steps in nuclear fission?

Advantages and Disadvantages of Nuclear Energy

One advantage of using nuclear energy is that a relatively small amount of uranium produces a large amount of energy. In addition, a well-run nuclear power plant does not pollute the air, the soil, or the water.

However, using nuclear energy has disadvantages. Nuclear power plants use a nonrenewable resource—uranium—for fuel. In addition, the chain reaction in the nuclear reactor must be carefully monitored. If it gets out of control, it can lead to a release of harmful radioactive substances into the environment.

The waste materials from nuclear power plants are highly radioactive and dangerous to living things. The waste materials remain dangerous for thousands of years. Storing them safely is important for both the environment and public health.

 Reading Check Why is it important to control a chain reaction?

Managing Nonrenewable Energy Resources

As shown in **Figure 6,** fossil fuels and nuclear energy provide about 93 percent of U.S. energy. Because these sources eventually will be gone, we must understand how to manage and conserve them. This is particularly important because energy use in the United States is higher than in other countries. Although only about 4.5 percent of the world's population lives in the United States, it uses more than 22 percent of the world's total energy.

Management Solutions

Mined land must be reclaimed. **Reclamation** *is a process in which mined land must be recovered with soil and replanted with vegetation.* Laws also help ensure that mining and drilling take place in an environmentally safe manner. In the United States, the Clean Air Act limits the amount of pollutants that can be released into the air. In addition, the U.S. Atomic Energy Act and the Energy Policy Act include **regulations** that protect people from nuclear emissions.

What You Can Do

Have you ever heard of vampire energy? Vampire energy is the energy used by appliances and other electronic equipment, such as microwave ovens, washing machines, televisions, and computers, that are plugged in 24 h a day. Even when turned off, they still consume energy. These appliances consume about 5 percent of the energy used each year. You can conserve energy by unplugging DVD players, printers, and other appliances when they are not in use.

You also can walk or ride your bike to help conserve energy. And, you can use renewable energy resources, which you will read about in the next lesson.

 Key Concept Check How can you help manage nonrenewable resources wisely?

Sources of Energy Used in the U.S. in 2007

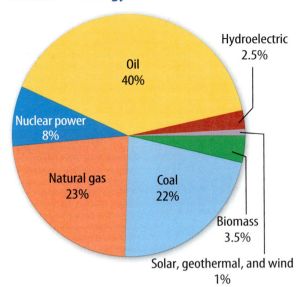

Figure 6 About 93 percent of the energy used in the United States comes from nonrenewable resources.

Visual Check Which energy source is used most in the United States?

ACADEMIC VOCABULARY
regulation
(noun) a rule dealing with procedures, such as safety

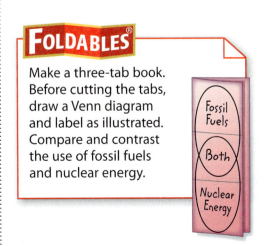

Make a three-tab book. Before cutting the tabs, draw a Venn diagram and label as illustrated. Compare and contrast the use of fossil fuels and nuclear energy.

Lesson 1 Review

 Assessment   Online Quiz

Visual Summary

Fossil fuels include coal, oil, and natural gas. Fossil fuels take millions of years to form. Humans use fossil fuels at a much faster rate.

Nuclear energy comes from splitting atoms, or fission. Nuclear power plants must be monitored for safety, and nuclear waste must be stored properly.

It is important to manage nonrenewable energy resources wisely. This includes mine reclamation, limiting air pollutants, and conserving energy.

FOLDABLES

Use your lesson Foldable to review the lesson. Save your Foldable for the project at the end of the chapter.

What do you think NOW?

You first read the statements below at the beginning of the chapter.

1. Nonrenewable energy resources include fossil fuels and uranium.
2. Energy use in the United States is lower than in other countries.

Did you change your mind about whether you agree or disagree with the statements? Rewrite any false statements to make them true.

Use Vocabulary

1. Energy produced from atomic reactions is called _____.
2. **Distinguish** between renewable and nonrenewable resources.
3. **Use the term** *reclamation* in a sentence.

Understand Key Concepts

4. What is the source of most energy in the United States?
   A. coal
   B. oil
   C. natural gas
   D. nuclear energy
5. **Summarize** the advantages and disadvantages of using nuclear energy.
6. **Illustrate** Make a poster showing how you can conserve energy.

Interpret Graphics

7. **Sequence** Draw a graphic organizer like the one below to sequence the events in the formation of oil.

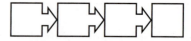

8. **Describe** Use the diagram below to describe the energy conversions that take place in a nuclear power plant.

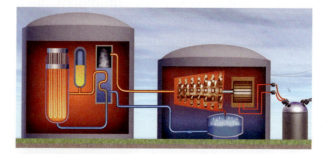

Critical Thinking

9. **Suppose** that a nuclear power plant will be built near your town. Would you support the plan? Why or why not?
10. **Consider** Do the advantages of using fossil fuels outweigh the disadvantages? Explain your answer.

Inquiry Skill Practice: Avoid Bias

30 minutes

How can you identify bias and its source?

Whenever an author is trying to persuade, or convince, readers to share a particular opinion, you must read and evaluate carefully for bias. Bias is a way of thinking that tells only one side of a story, sometimes with inaccurate information.

Learn It
Sometimes a scientific investigation involves making judgments. When you make a judgment, you form an opinion. It is important to be honest and not allow any expectations of results to **bias** your judgments.

Try It

1. Read the passage to the right for sources of bias, such as
   - claims not supported by evidence;
   - persuasive statements;
   - the author wanting to believe what he or she is saying, whether or not it is true.

Apply It

2. Analyze the passage, and identify two instances of bias and the source of each. Record this information in your Science Journal.

3. If you were the moderator at an EPA hearing about the issue, what would you do to solve the problem of bias?

4. **Key Concept** In your own words, explain how you would formulate an argument about the wise management of air resources while avoiding bias.

Factory Spews Out Air Pollution

Environmental organizations claim a coal-burning factory is polluting the air with toxic particulate matter in violation of U.S. Environmental Protection Agency (EPA) standards. Particulate matter is a mix of both solid and liquid particles in the air.

Citizens of the town collected air samples for a period of six months. An independent laboratory analysis of the samples showed a dangerously high level of the toxic particulate materials. Levels this high have been cited in medical journals as contributing to illness and death from asthma, respiratory disease, and lung cancer.

The factory has not updated its pollution-control equipment, claiming that it cannot afford the cost. At a town meeting, a company spokesperson claimed that the particulate matter is not harmful to human health. The state environmental director, who previously worked at the factory, stated that the jobs provided by the factory are more important to the state than environmental concerns.

Lesson 1
EXTEND
651

Lesson 2

Reading Guide

Key Concepts
ESSENTIAL QUESTIONS

- What are the main sources of renewable energy?
- What are the advantages and disadvantages of using renewable energy resources?
- What can individuals do to encourage the use of renewable energy resources?

Vocabulary
solar energy p. 653
wind farm p. 654
hydroelectric power p. 654
geothermal energy p. 655
biomass energy p. 655

 Multilingual eGlossary

Renewable Energy Resources

Inquiry What do these panels do?

These solar panels convert energy from the Sun into electrical energy. This solar power plant, at Nellis Air Force Base in Nevada, produces about 25 percent of the electricity used on the base. What are some of the advantages of using energy from the Sun? What are some of the disadvantages?

Launch Lab

20 minutes

How can renewable energy sources generate energy in your home?
Renewable energy technologies can contribute to reducing our dependence on fossil fuels.

1. Review the table below. It shows how much energy, in Watt-hours, it takes to run certain appliances.
2. In one hour, a typical bicycle generator generates 200 W-h of electric energy; a small solar panel generates 150 W-h; and small wind turbines typically generate 100 W-h. Complete the table by calculating the time it would take for each alternative form of energy to generate the electricity needed to run each appliance for 1 h.

Hint: Use the following equation to solve for the time used by each energy source:

$$\begin{pmatrix}\text{Time used by}\\\text{energy source}\end{pmatrix} = \frac{\begin{pmatrix}\text{Time to use}\\\text{appliance}\end{pmatrix} \times \begin{pmatrix}\text{Energy used per hour}\\\text{by appliance}\end{pmatrix}}{\begin{pmatrix}\text{Energy produced per hour}\\\text{by energy source}\end{pmatrix}}$$

Think About This
1. Which appliance required the longest energy-generating time from the alternative energy sources? Why?
2. 🔑 **Key Concept** What issues would you have to consider when using solar or wind energy to generate electricity in your home?

Appliance	Energy Used Per Hour	Time on Bike	Time for Solar Panel	Time for Wind Turbine
Desktop computer	75 W-h			
Hair dryer	1000 W-h			
Television	200 W-h			

Renewable Energy Resources

Could you stop the Sun from shining or the wind from blowing? These might seem like silly questions, but they help make an important point about renewable resources. Renewable resources come from natural processes that have been happening for billions of years and will continue to happen.

Solar Energy

Solar energy *is energy from the Sun.* Solar cells, such as those in watches and calculators, capture light energy and transform it to electrical energy. Solar power plants can generate electricity for large areas. They transform energy in sunlight, which then turns turbines connected to generators.

Some people use solar energy in their homes, as shown in **Figure 7.** Active solar energy uses technology, such as solar panels, that gathers and stores solar energy that heats water and homes. Passive solar energy uses design elements that capture energy in sunlight. For example, windows on the south side of a house can let in sunlight that helps heat a room.

Figure 7 🔑 People can use solar energy to provide electricity for their homes.

Wind Energy

Have you ever dropped your school papers outside and had them scattered by the wind? If so, you experienced wind energy. This renewable resource has been used since ancient times to sail boats and to turn windmills. Today, wind turbines, such as the ones shown in **Figure 8,** can produce electricity on a large scale. *A group of wind turbines that produce electricity is called a* **wind farm.**

✓ **Reading Check** How is wind energy a renewable resource?

Water Energy

Like wind energy, flowing water has been used as an energy source since ancient times. Today, water energy produces electricity using different methods, such as hydroelectric power and tidal power.

Hydroelectric Power *Electricity produced by flowing water is called* **hydroelectric power.** To produce hydroelectric power, humans build a dam across a powerful river. **Figure 9** shows how flowing water is used to produce electricity.

Tidal Power Coastal areas that have great differences between high and low tides can be a source of tidal power. Water flows across turbines as the tide comes in during high tides and as it goes out during low tides. The flowing water turns turbines connected to generators that produce electricity.

▲ **Figure 8** Offshore wind farms are called wind parks. This wind park is in Denmark.

Figure 9 In a hydroelectric power plant, energy from flowing water produces electricity. ▼

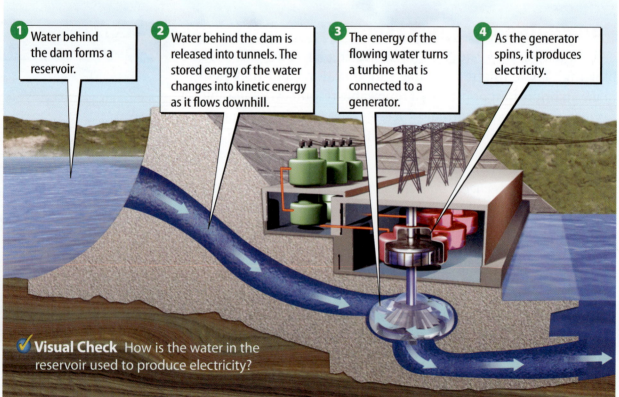

1. Water behind the dam forms a reservoir.
2. Water behind the dam is released into tunnels. The stored energy of the water changes into kinetic energy as it flows downhill.
3. The energy of the flowing water turns a turbine that is connected to a generator.
4. As the generator spins, it produces electricity.

✓ **Visual Check** How is the water in the reservoir used to produce electricity?

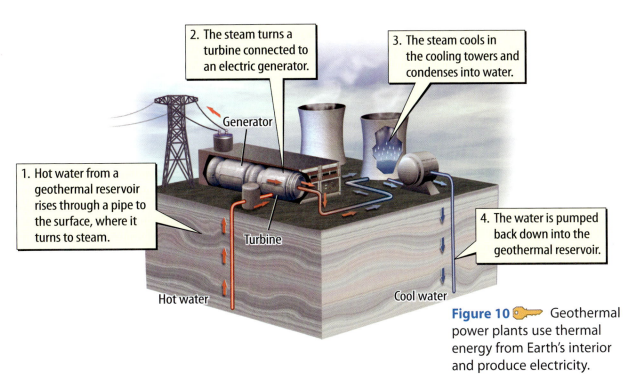

1. Hot water from a geothermal reservoir rises through a pipe to the surface, where it turns to steam.
2. The steam turns a turbine connected to an electric generator.
3. The steam cools in the cooling towers and condenses into water.
4. The water is pumped back down into the geothermal reservoir.

Figure 10 Geothermal power plants use thermal energy from Earth's interior and produce electricity.

Concepts in Motion
Animation

Geothermal Energy

Earth's core is nearly as hot as the Sun's surface. This thermal energy flows outward to Earth's surface. *Thermal energy from Earth's interior is called* **geothermal energy**. It can be used to heat homes and generate electricity in power plants, such as the one shown in **Figure 10.** People drill wells to reach hot, dry rocks or bodies of magma. The thermal energy from the hot rocks or magma heats water that makes steam. The steam turns turbines connected to generators that produce electricity.

WORD ORIGIN
geothermal
from Greek *ge-*, means "Earth"; and Greek *therme*, means "heat"

Biomass Energy

Since humans first lit fires for warmth and cooking, biomass has been an energy source. **Biomass energy** *is energy produced by burning organic matter, such as wood, food scraps, and alcohol.* Wood is the most widely used biomass. Industrial wood scraps and organic materials, such as grass clippings and food scraps, are burned to generate electricity on a large scale.

Biomass also can be converted into fuels for vehicles. Ethanol is made from sugars in plants, such as corn. Ethanol often is blended with gasoline. This reduces the amount of oil used to make the gasoline. Adding ethanol to gasoline also reduces the amount of carbon monoxide and other pollutants released by vehicles. Another renewable fuel, biodiesel, is made from vegetable oils and fats. It emits few pollutants and is the fastest-growing renewable fuel in the United States.

FOLDABLES
Make a vertical five-tab Foldable. Label the tabs as illustrated. Identify the advantages and disadvantages of alternative fuels.

Key Concept Check What are the main sources of renewable energy?

Lesson 2
EXPLAIN

Advantages and Disadvantages of Renewable Resources

A big advantage of using renewable energy resources is that they are renewable. They will be available for millions of years to come. In addition, renewable energy resources produce less pollution than fossil fuels.

There are disadvantages associated with using renewable resources, however. Some are costly or limited to certain areas. For example, large-scale geothermal plants are limited to areas with tectonic activity. Recall that tectonic activity involves the movement of Earth's plates. **Table 2** lists the advantages and disadvantages of using renewable energy resources.

Key Concept Check What are some advantages and disadvantages of using renewable energy resources?

Table 2 Most renewable energy resources produce little or no pollution.

 Visual Check What are the advantages and the disadvantages of biomass energy?

Table 2 Renewable Resources—Advantages and Disadvantages

Renewable Resource	Advantages	Disadvantages
Solar energy	• nonpolluting • available in the United States	• less energy produced on cloudy days • no energy produced at night • high cost of solar cells • requires a large surface area to collect and produce energy on a large scale
Wind energy	• nonpolluting • relatively inexpensive • available in the United States	• large-scale use limited to areas with strong, steady winds • best sites for wind farms are far from urban areas and transmission lines • potential impact on bird populations
Water energy	• nonpolluting • available in the United States	• large-scale use limited to areas with fast-flowing rivers or great tidal differences • negative impact on aquatic ecosystems • production of electricity affected by long periods of little or no rainfall
Geothermal energy	• produces little pollution • available in the United States	• large-scale use limited to tectonically active areas • habitat disruption from drilling to build a power plant
Biomass energy	• reduces amount of organic material discarded in landfills • available in the United States	• air pollution results from burning some forms of biomass • less energy efficient than fossil fuels, costly to transport

Managing Renewable Energy Resources

Renewable energy currently meets only 7 percent of U.S. energy needs. As shown in **Figure 11,** most renewable energy comes from biomass. Solar energy, wind energy, and geothermal energy meet only a small percentage of U.S. energy needs. However, some states are passing laws that require the state's power companies to produce a percentage of electricity using renewable resources. Management of renewable resources often focuses on encouraging their use.

Management Solutions

The U.S. government has begun programs to encourage use of renewable resources. In 2009, billions of dollars were granted to the U.S. Department of Energy's Office of Efficiency and Renewable Energy for renewable energy research and programs that reduce the use of fossil fuels.

What You Can Do

You might be too young to own a house or a car, but you can help educate others about renewable energy resources. You can talk with your family about ways to use renewable energy at home. You can participate in a renewable energy fair at school. As a consumer, you also can make a difference by buying products that are made using renewable energy resources.

 Key Concept Check What can you do to encourage the use of renewable energy resources?

Energy Resources in the United States

Nonrenewable energy resources 93%

Renewable energy resources 7%

Energy Resource	Percent
Biomass	53%
Hydroelectric	36%
Wind	5%
Geothermal	5%
Solar	1%

Figure 11 The renewable energy resource used most in the United States is biomass energy.

Inquiry MiniLab   20 minutes

How are renewable energy resources used at your school?

Complete a survey about the use of renewable resources in your school.

1. Prepare interview questions about the use of renewable energy resources at your school. Each group member should come up with at least two questions.
2. Choose one group member to interview a school staff member.
3. Copy the table at the right into your Science Journal, and fill in the interview data.

Renewable Energy Source	Yes/No	Where is it used?	Why is it used? or Why isn't it used?
Sun			
Wind			
Water			
Geothermal			
Biomass			

Analyze and Conclude

1. **Explain** Which renewable energy resources are and are not being used? Why or why not?
2. **Key Concept** Choose one "why not" reason and describe how it could be addressed by communication with school planners.

Lesson 2
EXPLAIN

Lesson 2 Review

Visual Summary

Renewable energy resources can be used to heat homes, produce electricity, and power vehicles.

Advantages of renewable energy resources include little or no pollution and availability.

Management of renewable energy resources includes encouraging their use and continuing to research more about their use.

FOLDABLES

Use your lesson Foldable to review the lesson. Save your Foldable for the project at the end of the chapter.

What do you think NOW?

You first read the statements below at the beginning of the chapter.

3. Renewable energy resources do not pollute the environment.
4. Burning organic material can produce electricity.

Did you change your mind about whether you agree or disagree with the statements? Rewrite any false statements to make them true.

Use Vocabulary

1. **Define** *hydroelectric power* in your own words.

2. Burning wood is an example of _____ energy.

Understand Key Concepts

3. Which can reduce the amount of organic material discarded in landfills?
   A. biomass energy   C. water energy
   B. solar energy   D. wind energy

4. **Compare and contrast** solar energy and wind energy.

5. **Determine** Your family wants to use renewable energy to heat your home. Which renewable energy resource is best suited to your area? Explain your answer.

Interpret Graphics

6. **Organize** Copy and fill in the graphic organizer below. In each oval, list a type of renewable energy resource.

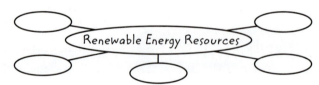

7. **Compare** the use of renewable resources and nonrenewable resources in the production of electricity in the United States, based on the table below.

Sources of Electricity Generation, 2007	
Energy Source	Percent
Fossil fuels	72.3
Nuclear power	19.4
Hydroelectric	5.8
Solar, wind, geothermal, biomass	2.5

Critical Thinking

8. **Design** and explain a model that shows how a renewable resource produces energy.

Inquiry Skill Practice — Analyze Data

40 minutes

How can you analyze energy-use data for information to help conserve energy?

As a student, you are not making large governmental policy decisions about uses of resources. As an individual, however, you can analyze data about energy use. You can use your analysis to determine some personal actions that can be taken to conserve energy resources.

Learn It

To **analyze the data** of fuel usage, you will need to look for patterns in the data, compare and categorize them, and determine cause and effect.

Try It

1. Study the fuel usage graph shown below. The data were collected from a house that uses natural gas as a source of energy to heat it.

2. Identify the time period that is covered by the graph.

3. Explain what is represented by the values on the vertical axis of the graph.

4. Describe the range of monthly gas usage over the 12-month period.

5. Group the monthly gas usage into three levels. Give each level a title. Enter these in your Science Journal.

Apply It

6. Categorize the three levels based on the amount of natural gas use.

7. Identify the three highest and four lowest months of gas usage. What might explain the usage patterns during these months?

8. Suppose the house from which the data came was heated with an electric furnace, instead of a furnace that used natural gas. What would you expect a usage graph for an electric furnace to look like?

9. **Key Concept** Formulate a list of heat conservation practices for homes.

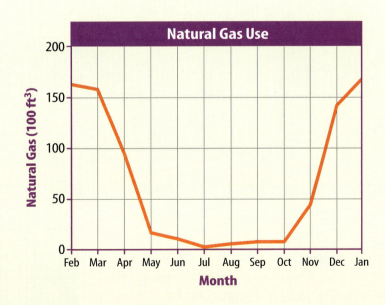

Lesson 2
EXTEND

Lesson 3

Land Resources

Reading Guide

Key Concepts
ESSENTIAL QUESTIONS

- Why is land considered a resource?
- What are the advantages and disadvantages of using land as a resource?
- How can individuals help manage land resources wisely?

Vocabulary

ore p. 663
deforestation p. 664

 Multilingual eGlossary

 Video

What's Science Got to do With It?

Inquiry A Garden on the Water?

The Science Barge is an experimental farm in New York City, New York. It saves space and reduces pollution and fossil fuel use while growing crops to feed people in an urban area. Why are people experimenting with ways to grow food that have fewer environmental impacts? Why is it important for humans to use land resources wisely?

Launch Lab

20 minutes

What resources from the land do you use every day?

The land on which humans live is part of Earth's crust. It provides resources that enable humans and other organisms to survive.

1. Make a list of every item you use in a 24-h period as you carry out your daily activities.
2. Combine your list with your group members' lists and decide which items contain resources from the land. Design a graphic organizer to group the materials into categories.
3. Fill in the graphic organizer on **chart paper.** Use a **highlighter** or **colored markers** to show which resources are renewable and which are nonrenewable.
4. Post your chart and compare it with the others in your class.

Think About This

1. Are there any times in your day when you do not use a resource from the land? Provide an example.
2. Describe the major categories that you used to organize your list of resources.
3. **Key Concept** Why do you think land is considered a resource?

Land as a Resource

A natural resource is something from Earth that living things use to meet their needs. People use soil for growing crops and forests to harvest wood for making furniture, houses, and paper products. They mine minerals from the land and clear large areas for roads and buildings. In each of these cases, people use land as a natural resource to meet their needs.

Key Concept Check Why is land considered a resource?

Living Space

No matter where you live, you and all living things use land for living space. Living space includes natural habitats, as well as the land on which buildings, sidewalks, parking lots, and streets are built. As shown in **Figure 12,** cities make up only a small percentage of land use in the United States. Most land is used for agriculture, grasslands, and forests.

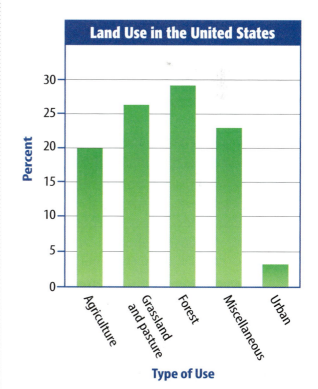

Figure 12 Forests and grasslands make up the largest categories of U.S. land use.

Lesson 3
EXPLORE
661

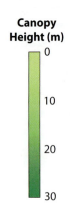

1650 **1992**

Figure 13 Much of the U.S. eastern forest has been replaced by cities, farms, and other types of development.

 Visual Check Compare forest cover in the eastern United States in 1650 and 1920.

Forests and Agriculture

As shown in **Figure 13**, forests covered much of the eastern United States in 1650. By 1920, the forests had nearly disappeared. Although some of the trees have grown back, they are not as tall and the forests are not as complex as they were originally.

Forests were cut down for the same reasons that forests are cut down today: for fuel, paper products, and wood products. People also cleared land for development and agriculture. Today, about one-fifth of U.S. land is used for growing crops and about one-fourth is used for grazing livestock.

 Reading Check Why are forests cut down?

Inquiry MiniLab

20 minutes

How can you manage land resource use with environmental responsibility?

You inherited a 100-acre parcel of forested land. Your relative's will stated that you must support all of your needs by using or selling resources from the land. To receive the inheritance, you must create an environmentally responsible land-use plan.

1. Copy the table into your Science Journal.
2. Your relative's will stated that you must support all of your needs by using or selling resources from the land. Decide how you will use the land, and complete the table.
3. Draw your land use plan on **graph paper**.
4. Present your group's land use plan to the class, and explain your reasoning.

Land Use	Percent of Total Area	Reasoning
Forest		
House and yard		
Garden		
Mineral mine		
Other		

Analyze and Conclude

1. **Compare and contrast** the design and reasoning of your plan with another group's plan.
2. **Identify** additional information about the land parcel that you would need to refine your plan.
3. 🔑 **Key Concept** Summarize two environmentally responsible practices that were used by more than one group in their plan.

Mineral Resources

Recall that coal, an energy resource, is mined from the land. Certain minerals also are mined to make products you use every day. These minerals often are called ores. **Ores** *are deposits of minerals that are large enough to be mined for a profit.*

The house in **Figure 14** contains many examples of common items made from mineral resources. Some of these come from metallic mineral resources. Ores such as bauxite and hematite are metallic mineral resources. They are used to make metal products. The aluminum in automobiles and refrigerators comes from bauxite. The iron in nails and faucets comes from hematite. Some mineral resources come from nonmetallic mineral resources, such as sand, gravel, gypsum, and halite. Nonmetallic mineral resources also are mined from the land. The sulfur used in paints and rubber and the fluorite used in paint pigments are other examples of nonmetal mineral resources.

WORD ORIGIN
ore
from Old English *ora*, means "unworked metal"

Figure 14 Many common products are made from mineral resources.

Visual Check Identify two products made from nonmetallic mineral resources.

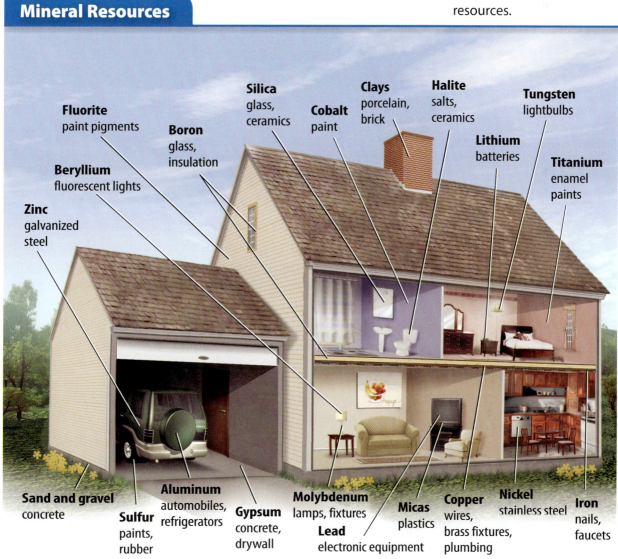

Mineral Resources

- **Fluorite** paint pigments
- **Beryllium** fluorescent lights
- **Zinc** galvanized steel
- **Boron** glass, insulation
- **Silica** glass, ceramics
- **Cobalt** paint
- **Clays** porcelain, brick
- **Halite** salts, ceramics
- **Tungsten** lightbulbs
- **Lithium** batteries
- **Titanium** enamel paints
- **Sand and gravel** concrete
- **Sulfur** paints, rubber
- **Aluminum** automobiles, refrigerators
- **Gypsum** concrete, drywall
- **Molybdenum** lamps, fixtures
- **Lead** electronic equipment
- **Micas** plastics
- **Copper** wires, brass fixtures, plumbing
- **Nickel** stainless steel
- **Iron** nails, faucets

FOLDABLES

Make a horizontal two-tab book. Label the tabs as illustrated. Use the Foldable to record your notes about renewable and nonrenewable land resources.

| Renewable Land Resources | Nonrenewable Land Resources |

Advantages and Disadvantages of Using Land Resources

Land resources such as soil and forests are widely available and easy to access. In addition, crops and trees are renewable—they can be replanted and grown in a relatively short amount of time. These are all advantages of using land resources.

Some land resources, however, are nonrenewable. It can take millions of years for minerals to form. This is one disadvantage of using land resources. Other disadvantages include deforestation and pollution.

Deforestation

As shown in **Figure 15,** humans sometimes cut forests to clear land for grazing, farming, and other uses. **Deforestation** *is the cutting of large areas of forests for human activities.* It leads to soil erosion and loss of animal habitats. In tropical rain forests—complex ecosystems that can take hundreds of years to replace—deforestation is a serious problem.

Figure 15 Deforestation occurs when humans cut forests to clear land for agricultural uses or development.

Deforestation also can affect global climates. Trees remove carbon dioxide from the atmosphere during photosynthesis. Rates of photosynthesis decrease when large areas of trees are cut down, and more carbon dioxide remains in the atmosphere. Carbon dioxide helps trap thermal energy within Earth's atmosphere. Increased concentrations of carbon dioxide can cause Earth's average surface temperatures to increase.

Pollution

Recall that **runoff** from coal mines can affect soil and water quality. The same is true of mineral mines. Runoff that contains chemicals from these mines can pollute soil and water. In addition, many farmers use chemical fertilizers to help grow crops. Runoff containing fertilizers can pollute rivers, soil, and underground water supplies.

 Key Concept Check What are some advantages and disadvantages of using land resources?

REVIEW VOCABULARY
runoff
rainwater that does not soak into the ground and flows over Earth's surface

Managing Land Resources

Because some land uses involve renewable resources while others do not, managing land resources is complex. For example, a tree is renewable. But forests can be nonrenewable because some can take hundreds of years to fully regrow. In addition, the amount of land is limited, so there is competition for space. Those who manage land resources must balance all of these issues.

Management Solutions

One way governments can manage forests and other unique ecosystems is by **preserving** them. On preserved land, logging and development is either banned or strictly controlled. Large areas of forests cannot be cut. Instead, loggers cut selected trees and then plant new trees to replace ones they cut.

Land mined for mineral resources also must be preserved. On both public and private lands, mined land must be restored according to government regulations.

Land used for farming and grazing can be managed to conserve soil and improve crop yield. Farmers can leave crop stalks after harvesting to protect soil from erosion. They also can use organic farming techniques that do not use synthetic fertilizers.

What You Can Do

You can help conserve land resources by recycling products made from land resources. You can use yard waste and vegetable scraps to make rich compost for gardening, reducing the need to use synthetic fertilizers. Compost is a mix of decayed organic material, bacteria and other organisms, and small amounts of water. **Figure 16** shows another way you can help manage land resources wisely.

 Key Concept Check What can you do to help manage land resources wisely?

> **SCIENCE USE V. COMMON USE**
>
> **preserve**
>
> ***Science Use*** to keep safe from injury, harm, or destruction
>
> ***Common Use*** to can, pickle, or save something for future use

Figure 16
A community garden is one way to help manage land resources wisely.

Lesson 3 Review

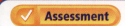 Assessment Online Quiz

Visual Summary

Land is a natural resource that humans use to meet their needs.

Disadvantages of using land as a resource include deforestation, which leads to increased erosion and increased carbon dioxide in the atmosphere.

Individuals can help manage land resources wisely by recycling, composting, and growing food in community gardens.

FOLDABLES

Use your lesson Foldable to review the lesson. Save your Foldable for the project at the end of the chapter.

What do you think NOW?

You first read the statements below at the beginning of the chapter.

5. Cities cover most of the land in the United States.
6. Minerals form over millions of years.

Did you change your mind about whether you agree or disagree with the statements? Rewrite any false statements to make them true.

Use Vocabulary

1. Cutting down forests for human activities is called _____.
2. **Use the word** *ore* in a sentence.

Understand Key Concepts

3. One disadvantage of using metallic mineral resources is that these resources are
   A. easy to mine.   C. nonrenewable.
   B. inexpensive.   D. renewable.

4. **Give an example** of how people use land as a resource.

5. **Compare** the methods used by governments and individuals to manage land resources wisely.

Interpret Graphics

6. **Take Notes** Copy the graphic organizer below, and list at least two land resources mentioned in this lesson. Describe how using each affects the environment.

Land Resource	How Use Affects Environment

7. **Identify** whether the mineral resources shown here are metallic or nonmetallic.

Critical Thinking

8. **Design** a way to manage land resources wisely. Use a method that is not discussed in this lesson.

9. **Decide** Land is a limited resource. There often is pressure to develop preserved land. Do you think this should happen? Why or why not?

A Greener Greensburg

GREEN SCIENCE

A town struck by disaster makes the world a greener place.

In May 2007, a powerful tornado struck the small Kansas town of Greensburg. The tornado destroyed almost every home, school, and business. Six months later, the town's officials and residents decided to rebuild Greensburg as a model green community.

The town's residents pledged to use fewer natural resources; to produce clean, renewable energy; and to reuse and recycle waste. As part of this effort, every new home and building would be designed for energy efficiency. The homes also would be constructed of materials that are healthful for the people who live and work in them.

What is a model green town? Here are some ways Greensburg will help the environment, save money, and make life better for its residents.

▲ Rain gardens help improve water quality by filtering pollutants from runoff.

USE RENEWABLE ENERGY

- **Produce clean energy** with renewable energy sources such as wind and sunlight. Wind turbines capture the abundant wind power of the Kansas plains.
- **Cut back on greenhouse gas emissions** with electric or hybrid city vehicles.

CONSERVE WATER

- **Capture runoff and rainwater** with landscape features such as rain gardens, bowl-shaped gardens designed to collect and absorb excess rainwater.
- **Use low-flow** faucets, shower heads, and toilets.

CREATE A HEALTHY ENVIRONMENT

- **Provide parks and green spaces** filled with native plants that need little water or care.
- **Create a "walkable community"** to encourage people to drive less and be more active, with a town center connected to neighborhoods by sidewalks and trails.

BUILD GREEN BUILDINGS

- **Design every home, school, and office** to use less energy and promote better health.
- **Make the most of natural daylight** for indoor lighting with many windows, which also can be opened for fresh air.
- **Use green materials** that are nontoxic and locally grown or made from recycled materials.

PROBLEM SOLVING With your group, choose one of Greensburg's projects. Make a plan describing how it could be implemented in your community and what its benefits would be.

Lesson 4

Reading Guide

Key Concepts
ESSENTIAL QUESTIONS

- Why is it important to manage air and water resources wisely?
- How can individuals help manage air and water resources wisely?

Vocabulary
photochemical smog p. 670
acid precipitation p. 670

 Multilingual eGlossary

 Video BrainPOP®

Air and Water Resources

Inquiry **Are these crop circles?**

No, this dotted landscape in Colorado is the result of circle irrigation. The fields are round because the irrigation equipment pivots from the center of the field and moves in a circle to water the crops. Crop irrigation accounts for about 34 percent of water used in the United States.

Launch Lab

20 minutes

How often do you use water each day?
In most places in the United States, people are fortunate to have an adequate supply of clean water. When you turn on the faucet, do you think about the value of water as a resource?

1. Prepare a two-column table to collect data on the number of times you use water in one day. Title the first column *Purpose* and the second column *Times Used*.
2. In the *Purpose* column, describe how you used the water, such as *Faucet, Toilet, Shower/Bath, Dishwasher, Laundry, Leaks,* and *Other*.
3. In the *Times Used* column, record and tally the total number of times you used water.
4. Calculate the percent that you use water for each category. Construct a circle graph showing the percentages of use in a day.

Think About This
1. For which purpose did you use water the most? The least?
2. **Key Concept** In which category, or categories, could you conserve water? How?

Importance of Air and Water

Using some natural resources, such as fossil fuels and minerals, makes life easier. You would miss them if they were gone, but you would still survive.

Air and water, on the other hand, are resources that you cannot live without. Most living things can survive only a few minutes without air. Oxygen from air helps your body provide energy for your cells.

Water also is needed for many life functions. As shown in **Figure 17,** water is the main component of blood. Water helps protect body tissues, helps maintain body temperature, and has a role in many chemical reactions, such as the digestion of food. In addition to drinking water, people use water for other purposes that you will learn about later in this lesson, including agriculture, transportation, and recreation.

Reading Check What are the functions of water in the human body?

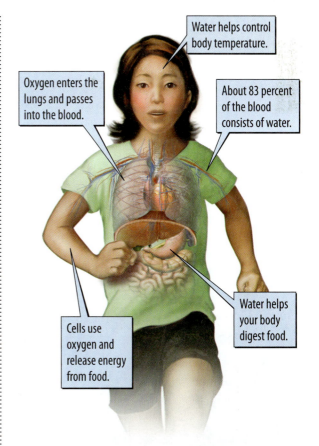

Figure 17 Your body needs oxygen and water to carry out its life-sustaining functions.

Figure 18 Sometimes a layer of warm air can trap smog in the cooler air close to Earth's surface. The smog can cover an area for days.

✅ **Visual Check** Where does the pollution that forms smog come from?

 Review   Personal Tutor

① During winter, the Sun's rays are less intense, so air near Earth's surface is cooler.

② Sometimes warmer air traps colder air and acts as a lid, holding cold air near the ground.

③ The warm air also traps a layer of pollution from vehicles, industry, and homes.

FOLDABLES

Make a horizontal two-tab book. Label it as illustrated. Use your Foldable to discuss the importance of air and water.

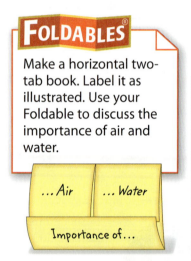

Air

Most living things need air to survive. However, polluted air, such as the air in **Figure 18,** can actually harm humans and other living things. Air pollution is produced when fossil fuels burn in homes, vehicles, and power plants. It also can be caused by natural events, such as volcanic eruptions or forest fires.

✅ **Reading Check** What activities can cause air pollution?

Smog Burning fossil fuels releases not only energy, but also substances, such as nitrogen compounds. **Photochemical smog** *is a brownish haze produced when nitrogen compounds and other pollutants in the air react in the presence of sunlight.* Smog can irritate your respiratory system. In some individuals, it can increase the chance of asthma attacks. Smog can be particularly harmful when it is trapped under a layer of warm air and remains in an area for several days, also shown in **Figure 18.**

Acid Precipitation Nitrogen and sulfur compounds released when fossil fuels burn can react with water in the atmosphere and produce acid precipitation. **Acid precipitation** *is precipitation that has a pH less than 5.6.* When it falls into lakes, it can harm fish and other organisms. It also can pollute soil and kill trees and other plants. Acid precipitation can even damage buildings and statues made of some types of rocks.

Figure 19 Gas and dust released by erupting volcanoes, such as Karymsky Volcano in Russia, can pollute the air.

Natural Events Forest fires and volcanic eruptions, such as the one shown in **Figure 19,** release gases, ash, and dust into the air. Dust and ash from one volcanic eruption can spread around the world. Materials from forest fires and volcanic eruptions can cause health problems similar to those caused by smog.

Water

Suppose you saved $100, but you were only allowed to spend 90 cents. You might be very frustrated! If all of the water on Earth were your $100, freshwater that we can use is like that 90 cents you can spend. As shown in **Figure 20**, most water on Earth is salt water. Only 3 percent is freshwater, and most of that is frozen in glaciers. That leaves just a small part, 0.9 percent, of the total amount of water on Earth for humans to use.

This relatively small supply of freshwater must meet many needs. In addition to drinking water, people use water for farming, industry, electricity production, household activities, transportation, and recreation. Each of these uses can affect water quality. For example, water used to irrigate fields can mix with fertilizers. This polluted water then can run off into rivers and groundwater, reducing the quality of these water supplies. Water used in industry often is heated to high temperatures. The hot water can harm aquatic organisms when it is returned to the environment.

 Reading Check How can farming affect water quality?

Water Distribution on Earth

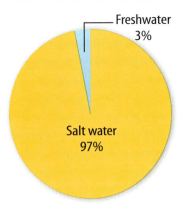

Figure 20 Freshwater makes up only 3 percent of Earth's water.

Inquiry MiniLab   20 minutes

How much water can a leaky faucet waste?

You are competing for the job of environmental consultant at your school. One of the competition requirements is to complete an analysis of water waste from existing faucets.

1. Read and complete a lab safety form.
2. Catch the water from a **leaking faucet** in a **beaker.** Time the collection for 1 min with a **stopwatch.**
3. Use a **50-mL graduated cylinder** to measure the amount of water lost. Record the amount of water, in milliliters per minute, that leaked from the faucet in your Science Journal.
4. Make a table to show the amount of water that would leak from the faucet in 1 hour, 1 day, 1 week, 1 month, and 1 year.

Analyze and Conclude

1. **Construct** a graph of your data. Label the axes and title your graph. Explain what the graph illustrates.
2. **Describe** how many liters of water would be wasted by the leak over a period of one year. Explain how you arrived at that figure.
3. **Key Concept** As an environmental consultant, what information and recommendations would your report contain about water waste in the school?

Lesson 4
EXPLAIN

Figure 21 The amount of sulfur compounds in the atmosphere decreased following the passage of the Clean Air Act.

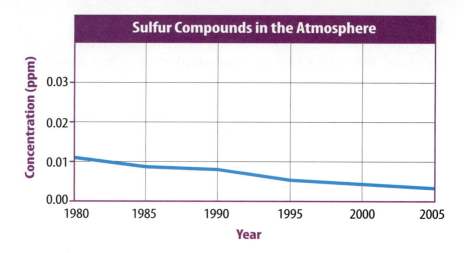

Math Skills

Use Percentages

The carbon monoxide (CO) level in Seattle air went from 7.8 parts per million (ppm) in 1990 to 1.8 ppm in 2007. What was the percent change in CO levels?

1. Subtract the starting value from the final value.

   1.8 ppm − 7.8 ppm = −6.0 ppm

2. Divide the difference by the starting value.

   −6.0 ppm / 7.8 ppm = −0.769

3. Multiply by 100 and add a % sign.

   −0.769 × 100 = −76.9%.

It decreased by 76.9%.

Practice

Between 1990 and 2000, the ozone (O_3) levels in New York City went from 0.098 ppm to 0.086 ppm. What was the percent change in ozone levels?

- Review
- Math Practice
- Personal Tutor

Managing Air and Water Resources

Animals and plants do not use natural resources to produce electricity or to raise crops. But they do use air and water. Management of these important resources must consider both human needs and the needs of other living things.

 Key Concept Check Why is it important to manage air and water resources wisely?

Management Solutions

Legislation is an effective way to reduce air and water pollution. The regulations of the U.S. Clean Air Act, passed in 1970, limit the amount of certain pollutants that can be released into the air. The graph in **Figure 21** shows how levels of sulfur compounds have decreased since the act became law.

Similar laws are now in place to maintain water quality. The U.S. Clean Water Act legislates the reduction of water pollution. The Safe Drinking Water Act legislates the protection of drinking water supplies. By reducing pollution, these laws help ensure that all living things have access to clean air and water.

What You Can Do

You have learned that reducing fossil fuel use and improving energy efficiency can reduce air pollution. You can make sure your home is energy efficient by cleaning air-conditioning or heating filters and using energy-saving lightbulbs.

You can help reduce water pollution by properly disposing of harmful chemicals so that less pollution runs off into rivers and streams. You can volunteer to help clean up litter from a local stream. You also can conserve water so there is enough of this resource for you and other living things in the future.

Key Concept Check How can individuals help manage air and water resources wisely?

Lesson 4 Review

Assessment — Online Quiz

Visual Summary

Sources of air pollution include the burning of fossil fuels in vehicles and power plants, and natural events such as volcanic eruptions and forest fires.

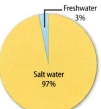

Only a small percentage of Earth's water is available for humans to use. Humans use water for agriculture, industry, recreation, and cleaning.

Management of air and water resources includes passing laws that regulate sources of air and water pollution. Individuals can reduce energy use and dispose of chemicals properly to help keep air and water clean.

Use your lesson Foldable to review the lesson. Save your Foldable for the project at the end of the chapter.

What do you think NOW?

You first read the statements below at the beginning of the chapter.

7. Humans need oxygen and water to survive.

8. About 10 percent of Earth's total water can be used by humans.

Did you change your mind about whether you agree or disagree with the statements? Rewrite any false statements to make them true.

Use Vocabulary

1. **Define** *acid precipitation* in your own words.

2. Air pollution caused by the reaction of nitrogen compounds and other pollutants in the presence of sunlight is _____.

Understand Key Concepts

3. About how much of Earth's water is available for humans to use?
   A. 0.01 percent   C. 3.0 percent
   B. 0.90 percent   D. 97.0 percent

4. **Relate** In terms of human health, why is it important to manage air resources wisely?

5. **List** ways your classroom could improve its energy efficiency.

Interpret Graphics

6. **Determine Cause and Effect** Copy and fill in the graphic organizer below to describe three effects of acid precipitation.

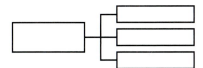

Critical Thinking

7. **Evaluate** The top three categories of household water use in the United States are flushing the toilet, washing clothes, and taking showers. Evaluate your water use, and list one thing you could do to reduce your use in each category.

Math Skills — Math Practice

8. Between 1990 and 2007, the amount of sulfur dioxide (SO_2) in Miami's air went from 0.0073 ppm to 0.0027 ppm. What was the percent change of SO_2?

Research Efficient Energy and Resource Use

1–3 class periods

A community organization is encouraging your school's board of education to participate in the "Green Schools" program. Your class has been nominated to research and report on the present status of energy efficiency and resource use in the school. The results of the report will be used as information for the presentation. Your task is to choose a natural resource and collect data about how it is presently used in the school. Your group will then recommend environmentally responsible management practices.

Question
How can a natural resource be used more wisely at school?

Procedure

1. Read and complete a lab safety form.
2. With your group, choose one of these resources to research its use in your school: water, land, air, or an energy resource.
3. For your chosen resource, plan how you will research resource use. What questions will you ask? How much of the resource is used by the school? Is it used efficiently? How could it be used more efficiently, or how could it be conserved? Have your teacher approve your plan.
4. Prepare data collection forms like the one below to record the results of your research in your Science Journal.
5. Conduct your research, and enter the data on the forms.

Sample Data Table				
Resource: Water				
Areas of Research: Water Loss Through Leaks and Recycling System				
Location	Faucets	Water Fountains	Toilets	"Gray" Water Recycling System
Washroom	6 good 2 poor 2 leaking		4 good 1 leaking	no
Hallway		3 good 1 leaking		no
Classroom 101	1 good			no
Classroom 102	1 good			no
Classroom 103	1 leaking			no

674 Chapter 18
EXTEND

6. Review and summarize the data. Perform any necessary calculations to convert values to annual usage.

7. Conduct interviews, or collect more data about areas of research for which you need additional information.

8. After analyzing your data, write a proposal suggesting how the resource can be wisely managed in your school.

9. Compare the elements you addressed in your research with those recommended by a state or a national environmental organization. Did your research include everything?

10. Modify your proposal, if necessary. Record your revisions in your Science Journal.

Analyze and Conclude

11. **Graph** and explain the results of your data analysis.

12. **Predict** one impact on the environment of the existing management practices of the resource that you audited.

13. **The Big Idea** Describe two recommendations that you would make to the school's board of education about changes in resource management practices.

Communicate Your Results

Present the results of your research and your proposal to the class. Use appropriate visual aids to help make your points.

 Extension

Combine information and reports from groups that investigated other resources from the list in step 2 so that all four resources are represented. Make a final report that includes recommendations for efficient use of each resource at your school.

Remember to use scientific methods.

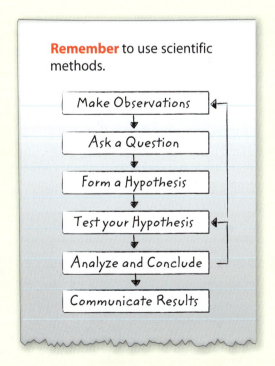

Lesson 4
EXTEND
675

Chapter 18 Study Guide

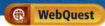

 Wise management of natural resources helps extend the supply of nonrenewable resources, reduce pollution, and improve soil, air, and water quality.

Key Concepts Summary

Vocabulary

Lesson 1: Energy Resources

- **Nonrenewable resources** include fossil fuels and uranium, which is used for **nuclear energy**.
- Nonrenewable energy resources are widely available and easy to convert to energy. However, using these resources can cause pollution and habitat disruption. Safety concerns also are an issue.
- People can conserve energy to help manage these resources.

nonrenewable resource p. 643
renewable resource p. 643
nuclear energy p. 647
reclamation p. 649

Lesson 2: Renewable Energy Resources

- Renewable energy resources include **solar energy**, wind energy, water energy, **geothermal energy**, and **biomass energy**.
- Renewable resources cause little to no pollution. However, some types of renewable energy are costly or limited to certain areas.
- Individuals can help educate others about renewable resources.

solar energy p. 653
wind farm p. 654
hydroelectric power p. 654
geothermal energy p. 655
biomass energy p. 655

Lesson 3: Land Resources

- Land is considered a resource because it is used by living things to meet their needs for food, shelter, and other things.
- Some land resources are renewable, while others are not.
- Individuals can recycle and compost to help conserve land resources.

ore p. 663
deforestation p. 664

Lesson 4: Air and Water Resources

- Most living things cannot survive without clean air and water.
- Individuals can make their homes and schools more energy efficient.

photochemical smog p. 670
acid precipitation p. 670

Study Guide

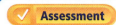
- **Personal Tutor**
- **Vocabulary eGames**
- **Vocabulary eFlashcards**

Chapter Project

Assemble your lesson Foldables as shown to make a Chapter Project. Use the project to review what you have learned in this chapter.

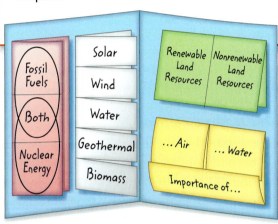

Use Vocabulary

1. Distinguish between renewable resources and nonrenewable resources.

2. Replace the underlined words with the correct vocabulary word: <u>Energy produced from atomic reactions</u> can be used to generate electricity.

3. How does biomass energy differ from geothermal energy?

4. Energy from the Sun is _____.

5. Define the term *ore* in your own words.

6. Distinguish between photochemical smog and acid precipitation.

Link Vocabulary and Key Concepts

Concepts in Motion — Interactive Concept Map

Copy these concept maps, and then use vocabulary terms from the previous page and other terms from the chapter to complete the concept maps.

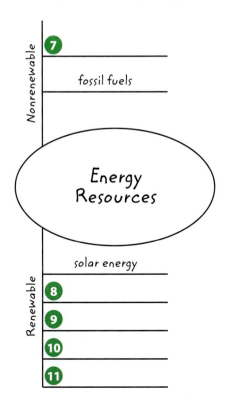
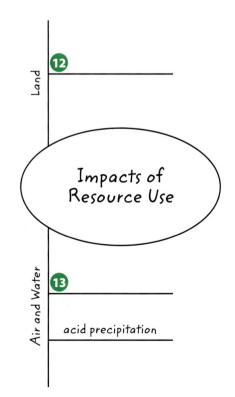

Chapter 18 Study Guide • **677**

Chapter 18 Review

Understand Key Concepts

1. Which energy source produces radioactive waste?
   A. biomass
   B. geothermal
   C. hydroelectric power
   D. nuclear power

2. The table below shows the energy sources used to produce electricity in the United States. What can you infer from the table?

Electricity Production	
Energy Source	Percent
Coal	48.5
Natural gas	21.6
Nuclear power	19.4
Hydroelectric power	5.8
Solar, wind, geothermal, biomass	2.5
Oil	1.6
Other	0.6

   A. About 19.4 percent of U.S. electricity comes from renewable sources.
   B. Hydroelectric power is more widely used for electricity than nuclear power.
   C. More than 90 percent of U.S. electricity comes from nonrenewable sources.
   D. Oil is more widely used for electricity than hydroelectric power.

3. Which factor would best determine whether a home is suitable for solar energy?
   A. difference in tidal heights
   B. strength of daily winds
   C. nearness to tectonically active areas
   D. number of sunny days per year

4. Which product comes from a metallic mineral resource?
   A. aluminum
   B. drywall
   C. gravel
   D. table salt

5. Which is a renewable land resource?
   A. forests
   B. minerals
   C. soil
   D. trees

6. Where is most water on Earth located?
   A. lakes
   B. oceans
   C. rivers
   D. underground

7. Which natural event can result in air pollution?
   A. burning fossil fuels
   B. littering a stream
   C. runoff from farms
   D. volcanic eruption

8. The graph below shows how the amount of sulfur compounds in the atmosphere has changed since the passage of the Clean Air Act. Based on the data in the graph, what can you infer about the act?

   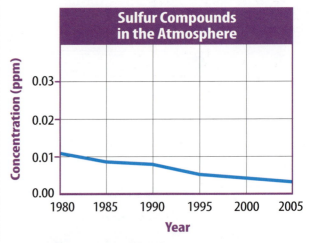

   A. The act has helped decrease pollutants in the atmosphere.
   B. The act has helped increase pollutants in the atmosphere.
   C. The act has incentives for use of renewable resources.
   D. The act has not impacted the amount of pollutants in the atmosphere.

Chapter Review

Assessment — Online Test Practice

Critical Thinking

9 Organize the list of energy sources into renewable and nonrenewable energy resources.

• coal	• nuclear energy
• solar energy	• wind energy
• oil	• natural gas
• geothermal energy	• tidal power
• hydroelectric power	• biomass

10 Create a cartoon showing a chain reaction in a nuclear power plant.

11 Compare hydroelectric and tidal power.

12 Design a way to use passive solar energy in your classroom.

13 Distinguish between geothermal energy and solar energy.

14 Consider What factors must governments consider when managing land resources?

15 Evaluate the use of forests as natural resources. Do the advantages outweigh the disadvantages? Explain.

16 Infer When would you expect more smog to form—on cloudy days or on sunny days? Explain.

17 Design a way to remove salt from salt water. Then evaluate your plan. Could it be used to produce freshwater on a large scale? Why or why not?

18 Formulate a way to demonstrate the importance of air and water resources to younger students.

Writing in Science

19 Compose a song about vampire energy. The lyrics should describe vampire energy and explain how it can be reduced.

REVIEW THE BIG IDEA

20 Select a natural resource and explain why it is important to manage the resource wisely.

21 Suppose the house below is heated by electricity produced from burning coal. Which areas of the house have the greatest loss of thermal energy? Why is it important for this house to reduce thermal energy loss?

Math Skills

Review — Math Practice

Use Percentages

22 Between 2002 and 2003, the carbon monoxide level in the air in Denver, Colorado, went from 2.9 ppm to 3.3 ppm. What was the percent change in CO?

23 There often is a considerable difference between pollutants in surface water and pollutants in groundwater in the same area. For example, in Portland, Oregon, there were 4.6 ppm of sulfates in the groundwater and 0.9 ppm in the surface water. What was the percent difference? (Hint: Use 4.6 ppm as the starting value.)

Standardized Test Practice

Record your answers on the answer sheet provided by your teacher or on a sheet of paper.

Multiple Choice

1. Which activity does NOT reduce the use of fossil fuels?
   A riding a bicycle to school
   B unplugging DVD players
   C walking to the store
   D watering plants less often

Use the graph below to answer questions 2 and 3.

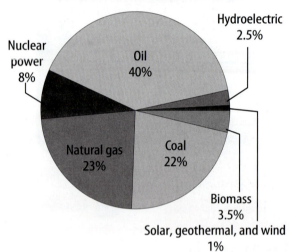

2. Which is the most-used renewable energy resource in the United States?
   A biomass
   B hydroelectric
   C natural gas
   D nuclear energy

3. What percentage of the energy used in the United States comes from burning fossil fuels?
   A 40 percent
   B 45 percent
   C 85 percent
   D 93 percent

4. Which practice emphasizes the use of renewable energy resources?
   A buying battery-operated electronics
   B installing solar panels on buildings
   C replacing sprinklers with watering cans
   D teaching others about vampire energy

5. Which is a nonrenewable land resource?
   A crops
   B minerals
   C streams
   D trees

Use the figure below to answer question 6.

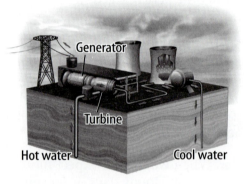

6. Which alternative energy resource is used to make electricity in the figure?
   A solar energy
   B tidal power
   C geothermal energy
   D hydroelectric power

7. Which practice is a wise use of land resources?
   A composting
   B conserving water
   C deforestation
   D strip mining

Standardized Test Practice

Assessment
Online Standardized Test Practice

Use the figure below to answer question 8.

8. Which type of air pollution is labeled *A* in the figure?
   - **A** acid precipitation
   - **B** fertilizer runoff
   - **C** nuclear waste
   - **D** photochemical smog

9. Approximately how much water on the Earth is in oceans?
   - **A** 1 percent
   - **B** 3 percent
   - **C** 75 percent
   - **D** 97 percent

10. Which is a source of biomass energy?
    - **A** sunlight
    - **B** uranium
    - **C** wind
    - **D** wood

Constructed Response

Use the figure below to answer questions 11 and 12.

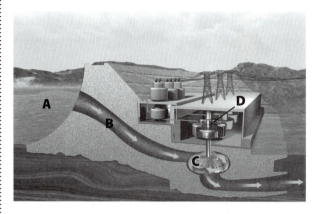

11. Which resource powers the turbine in the figure? Describe what happens at steps A–D to produce electricity.

12. What are two advantages and two disadvantages of producing electricity in the way shown in the figure?

13. Describe an example of how forests are used as a resource. What is one advantage of using the resource in this way? What is a disadvantage?

14. Agree or disagree with the following statement: "Known oil reserves will last only another 50 years. Thus, the United States should build more nuclear power plants to deal with the upcoming energy shortage." Support your answer with at least two advantages or two disadvantages of using nuclear energy.

NEED EXTRA HELP?														
If You Missed Question...	1	2	3	4	5	6	7	8	9	10	11	12	13	14
Go to Lesson...	1	2	1	2	3	2	3	4	4	2	2	2	3	1

Chapter 18 Standardized Test Practice • **681**

Student Resources

For Students and Parents/Guardians

These resources are designed to help you achieve success in science. You will find useful information on laboratory safety, math skills, and science skills. In addition, science reference materials are found in the Reference Handbook. You'll find the information you need to learn and sharpen your skills in these resources.

Table of Contents

Science Skill Handbook SR-2

Scientific Methods SR-2
- Identify a Question.. SR-2
- Gather and Organize Information......................... SR-2
- Form a Hypothesis.. SR-5
- Test the Hypothesis SR-6
- Collect Data SR-6
- Analyze the Data... SR-9
- Draw Conclusions SR-10
- Communicate... SR-10

Safety Symbols SR-11

Safety in the Science Laboratory SR-12
- General Safety Rules SR-12
- Prevent Accidents.. SR-12
- Laboratory Work SR-13
- Emergencies.. SR-13

Math Skill Handbook SR-14

Math Review... SR-14
- Use Fractions SR-14
- Use Ratios SR-17
- Use Decimals SR-17
- Use Proportions.. SR-18
- Use Percentages SR-19
- Solve One-Step Equations.................................... SR-19
- Use Statistics... SR-20
- Use Geometry SR-21

Science Application SR-24
- Measure in SI SR-24
- Dimensional Analysis SR-24
- Precision and Significant Digits SR-26
- Scientific Notation SR-26
- Make and Use Graphs SR-27

Foldables Handbook SR-29

Reference Handbook SR-40
- Periodic Table of the Elements............................. SR-40
- Topographic Map Symbols SR-42
- Rocks.. SR-43
- Minerals SR-44
- Weather Map Symbols SR-46

Glossary G-2

Index I-2

Credits C-2

Scientific Methods

Scientists use an orderly approach called the scientific method to solve problems. This includes organizing and recording data so others can understand them. Scientists use many variations in this method when they solve problems.

Identify a Question

The first step in a scientific investigation or experiment is to identify a question to be answered or a problem to be solved. For example, you might ask which gasoline is the most efficient.

Gather and Organize Information

After you have identified your question, begin gathering and organizing information. There are many ways to gather information, such as researching in a library, interviewing those knowledgeable about the subject, and testing and working in the laboratory and field. Fieldwork is investigations and observations done outside of a laboratory.

Researching Information Before moving in a new direction, it is important to gather the information that already is known about the subject. Start by asking yourself questions to determine exactly what you need to know. Then you will look for the information in various reference sources, like the student is doing in **Figure 1.** Some sources may include textbooks, encyclopedias, government documents, professional journals, science magazines, and the Internet. Always list the sources of your information.

Figure 1 The Internet can be a valuable research tool.

Evaluate Sources of Information Not all sources of information are reliable. You should evaluate all of your sources of information, and use only those you know to be dependable. For example, if you are researching ways to make homes more energy efficient, a site written by the U.S. Department of Energy would be more reliable than a site written by a company that is trying to sell a new type of weatherproofing material. Also, remember that research always is changing. Consult the most current resources available to you. For example, a 1985 resource about saving energy would not reflect the most recent findings.

Sometimes scientists use data that they did not collect themselves, or conclusions drawn by other researchers. This data must be evaluated carefully. Ask questions about how the data were obtained, if the investigation was carried out properly, and if it has been duplicated exactly with the same results. Would you reach the same conclusion from the data? Only when you have confidence in the data can you believe it is true and feel comfortable using it.

Interpret Scientific Illustrations As you research a topic in science, you will see drawings, diagrams, and photographs to help you understand what you read. Some illustrations are included to help you understand an idea that you can't see easily by yourself, like the tiny particles in an atom in **Figure 2**. A drawing helps many people to remember details more easily and provides examples that clarify difficult concepts or give additional information about the topic you are studying. Most illustrations have labels or a caption to identify or to provide more information.

Network Tree A type of concept map that not only shows a relationship, but how the concepts are related is a network tree, shown in **Figure 3**. In a network tree, the words are written in the ovals, while the description of the type of relationship is written across the connecting lines.

When constructing a network tree, write down the topic and all major topics on separate pieces of paper or notecards. Then arrange them in order from general to specific. Branch the related concepts from the major concept and describe the relationship on the connecting line. Continue to more specific concepts until finished.

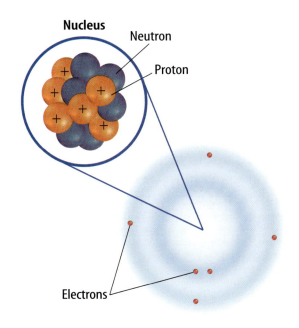

Figure 2 This drawing shows an atom of carbon with its six protons, six neutrons, and six electrons.

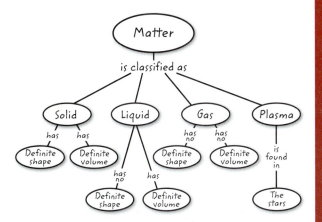

Figure 3 A network tree shows how concepts or objects are related.

Concept Maps One way to organize data is to draw a diagram that shows relationships among ideas (or concepts). A concept map can help make the meanings of ideas and terms more clear, and help you understand and remember what you are studying. Concept maps are useful for breaking large concepts down into smaller parts, making learning easier.

Events Chain Another type of concept map is an events chain. Sometimes called a flow chart, it models the order or sequence of items. An events chain can be used to describe a sequence of events, the steps in a procedure, or the stages of a process.

When making an events chain, first find the one event that starts the chain. This event is called the initiating event. Then, find the next event and continue until the outcome is reached, as shown in **Figure 4** on the next page.

Science Skill Handbook • **SR-3**

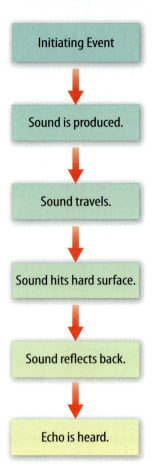

Figure 4 Events-chain concept maps show the order of steps in a process or event. This concept map shows how a sound makes an echo.

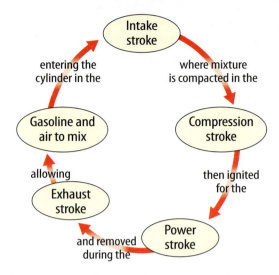

Figure 5 A cycle map shows events that occur in a cycle.

Spider Map A type of concept map that you can use for brainstorming is the spider map. When you have a central idea, you might find that you have a jumble of ideas that relate to it but are not necessarily clearly related to each other. The spider map on sound in **Figure 6** shows that if you write these ideas outside the main concept, then you can begin to separate and group unrelated terms so they become more useful.

Cycle Map A specific type of events chain is a cycle map. It is used when the series of events do not produce a final outcome, but instead relate back to the beginning event, such as in **Figure 5**. Therefore, the cycle repeats itself.

To make a cycle map, first decide what event is the beginning event. This is also called the initiating event. Then list the next events in the order that they occur, with the last event relating back to the initiating event. Words can be written between the events that describe what happens from one event to the next. The number of events in a cycle map can vary, but usually contain three or more events.

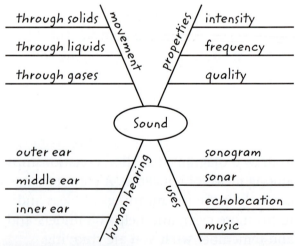

Figure 6 A spider map allows you to list ideas that relate to a central topic but not necessarily to one another.

SR-4 • Science Skill Handbook

Analyze the Data

To determine the meaning of your observations and investigation results, you will need to look for patterns in the data. Then you must think critically to determine what the data mean. Scientists use several approaches when they analyze the data they have collected and recorded. Each approach is useful for identifying specific patterns.

Interpret Data The word *interpret* means "to explain the meaning of something." When analyzing data from an experiment, try to find out what the data show. Identify the control group and the test group to see whether changes in the independent variable have had an effect. Look for differences in the dependent variable between the control and test groups.

Figure 14 A thermometer measures the temperature of an object.

Scientists use a thermometer to measure temperature. Most thermometers in a laboratory are glass tubes with a bulb at the bottom end containing a liquid such as colored alcohol. The liquid rises or falls with a change in temperature. To read a glass thermometer like the thermometer in **Figure 14,** rotate it slowly until a red line appears. Read the temperature where the red line ends.

Form Operational Definitions An operational definition defines an object by how it functions, works, or behaves. For example, when you are playing hide and seek and a tree is home base, you have created an operational definition for a tree.

Objects can have more than one operational definition. For example, a ruler can be defined as a tool that measures the length of an object (how it is used). It can also be a tool with a series of marks used as a standard when measuring (how it works).

Classify Sorting objects or events into groups based on common features is called classifying. When classifying, first observe the objects or events to be classified. Then select one feature that is shared by some members in the group, but not by all. Place those members that share that feature in a subgroup. You can classify members into smaller and smaller subgroups based on characteristics. Remember that when you classify, you are grouping objects or events for a purpose. Keep your purpose in mind as you select the features to form groups and subgroups.

Compare and Contrast Observations can be analyzed by noting the similarities and differences between two or more objects or events that you observe. When you look at objects or events to see how they are similar, you are comparing them. Contrasting is looking for differences in objects or events.

Science Skill Handbook • **SR-9**

Recognize Cause and Effect A cause is a reason for an action or condition. The effect is that action or condition. When two events happen together, it is not necessarily true that one event caused the other. Scientists must design a controlled investigation to recognize the exact cause and effect.

Draw Conclusions

When scientists have analyzed the data they collected, they proceed to draw conclusions about the data. These conclusions are sometimes stated in words similar to the hypothesis that you formed earlier. They may confirm a hypothesis, or lead you to a new hypothesis.

Infer Scientists often make inferences based on their observations. An inference is an attempt to explain observations or to indicate a cause. An inference is not a fact, but a logical conclusion that needs further investigation. For example, you may infer that a fire has caused smoke. Until you investigate, however, you do not know for sure.

Apply When you draw a conclusion, you must apply those conclusions to determine whether the data supports the hypothesis. If your data do not support your hypothesis, it does not mean that the hypothesis is wrong. It means only that the result of the investigation did not support the hypothesis. Maybe the experiment needs to be redesigned, or some of the initial observations on which the hypothesis was based were incomplete or biased. Perhaps more observation or research is needed to refine your hypothesis. A successful investigation does not always come out the way you originally predicted.

Avoid Bias Sometimes a scientific investigation involves making judgments. When you make a judgment, you form an opinion. It is important to be honest and not to allow any expectations of results to bias your judgments. This is important throughout the entire investigation, from researching to collecting data to drawing conclusions.

Communicate

The communication of ideas is an important part of the work of scientists. A discovery that is not reported will not advance the scientific community's understanding or knowledge. Communication among scientists also is important as a way of improving their investigations.

Scientists communicate in many ways, from writing articles in journals and magazines that explain their investigations and experiments, to announcing important discoveries on television and radio. Scientists also share ideas with colleagues on the Internet or present them as lectures, like the student is doing in **Figure 15.**

Figure 15 A student communicates to his peers about his investigation.

These safety symbols are used in laboratory and field investigations in this book to indicate possible hazards. Learn the meaning of each symbol and refer to this page often. *Remember to wash your hands thoroughly after completing lab procedures.*

PROTECTIVE EQUIPMENT Do not begin any lab without the proper protection equipment.

 GOGGLES Proper eye protection must be worn when performing or observing science activities that involve items or conditions as listed below.

 APRON Wear an approved apron when using substances that could stain, wet, or destroy cloth.

 SOAP Wash hands with soap and water before removing goggles and after all lab activities.

 GLOVES Wear gloves when working with biological materials, chemicals, animals, or materials that can stain or irritate hands.

LABORATORY HAZARDS

Symbols	Potential Hazards	Precaution	Response
DISPOSAL	contamination of classroom or environment due to improper disposal of materials such as chemicals and live specimens	• DO NOT dispose of hazardous materials in the sink or trash can. • Dispose of wastes as directed by your teacher.	• If hazardous materials are disposed of improperly, notify your teacher immediately.
EXTREME TEMPERATURE	skin burns due to extremely hot or cold materials such as hot glass, liquids, or metals; liquid nitrogen; dry ice	• Use proper protective equipment, such as hot mitts and/or tongs, when handling objects with extreme temperatures.	• If injury occurs, notify your teacher immediately.
SHARP OBJECTS	punctures or cuts from sharp objects such as razor blades, pins, scalpels, and broken glass	• Handle glassware carefully to avoid breakage. • Walk with sharp objects pointed downward, away from you and others.	• If broken glass or injury occurs, notify your teacher immediately.
ELECTRICAL	electric shock or skin burn due to improper grounding, short circuits, liquid spills, or exposed wires	• Check condition of wires and apparatus for fraying or uninsulated wires, and broken or cracked equipment. • Use only GFCI-protected outlets	• DO NOT attempt to fix electrical problems. Notify your teacher immediately.
CHEMICAL	skin irritation or burns, breathing difficulty, and/or poisoning due to touching, swallowing, or inhalation of chemicals such as acids, bases, bleach, metal compounds, iodine, poinsettias, pollen, ammonia, acetone, nail polish remover, heated chemicals, mothballs, and any other chemicals labeled or known to be dangerous	• Wear proper protective equipment such as goggles, apron, and gloves when using chemicals. • Ensure proper room ventilation or use a fume hood when using materials that produce fumes. • NEVER smell fumes directly. • NEVER taste or eat any material in the laboratory.	• If contact occurs, immediately flush affected area with water and notify your teacher. • If a spill occurs, leave the area immediately and notify your teacher.
FLAMMABLE	unexpected fire due to liquids or gases that ignite easily such as rubbing alcohol	• Avoid open flames, sparks, or heat when flammable liquids are present.	• If a fire occurs, leave the area immediately and notify your teacher.
OPEN FLAME	burns or fire due to open flame from matches, Bunsen burners, or burning materials	• Tie back loose hair and clothing. • Keep flame away from all materials. • Follow teacher instructions when lighting and extinguishing flames. • Use proper protection, such as hot mitts or tongs, when handling hot objects.	• If a fire occurs, leave the area immediately and notify your teacher.
ANIMAL SAFETY	injury to or from laboratory animals	• Wear proper protective equipment such as gloves, apron, and goggles when working with animals. • Wash hands after handling animals.	• If injury occurs, notify your teacher immediately.
BIOLOGICAL	infection or adverse reaction due to contact with organisms such as bacteria, fungi, and biological materials such as blood, animal or plant materials	• Wear proper protective equipment such as gloves, goggles, and apron when working with biological materials. • Avoid skin contact with an organism or any part of the organism. • Wash hands after handling organisms.	• If contact occurs, wash the affected area and notify your teacher immediately.
FUME	breathing difficulties from inhalation of fumes from substances such as ammonia, acetone, nail polish remover, heated chemicals, and mothballs	• Wear goggles, apron, and gloves. • Ensure proper room ventilation or use a fume hood when using substances that produce fumes. • NEVER smell fumes directly.	• If a spill occurs, leave area and notify your teacher immediately.
IRRITANT	irritation of skin, mucous membranes, or respiratory tract due to materials such as acids, bases, bleach, pollen, mothballs, steel wool, and potassium permanganate	• Wear goggles, apron, and gloves. • Wear a dust mask to protect against fine particles.	• If skin contact occurs, immediately flush the affected area with water and notify your teacher.
RADIOACTIVE	excessive exposure from alpha, beta, and gamma particles	• Remove gloves and wash hands with soap and water before removing remainder of protective equipment.	• If cracks or holes are found in the container, notify your teacher immediately.

Science Skill Handbook • **SR-11**

Safety in the Science Laboratory

Introduction to Science Safety

The science laboratory is a safe place to work if you follow standard safety procedures. Being responsible for your own safety helps to make the entire laboratory a safer place for everyone. When performing any lab, read and apply the caution statements and safety symbol listed at the beginning of the lab.

General Safety Rules

1. Complete the *Lab Safety Form* or other safety contract BEFORE starting any science lab.
2. Study the procedure. Ask your teacher any questions. Be sure you understand safety symbols shown on the page.
3. Notify your teacher about allergies or other health conditions that can affect your participation in a lab.
4. Learn and follow use and safety procedures for your equipment. If unsure, ask your teacher.

5. Never eat, drink, chew gum, apply cosmetics, or do any personal grooming in the lab. Never use lab glassware as food or drink containers. Keep your hands away from your face and mouth.
6. Know the location and proper use of the safety shower, eye wash, fire blanket, and fire alarm.

Prevent Accidents

1. Use the safety equipment provided to you. Goggles and a safety apron should be worn during investigations.
2. Do NOT use hair spray, mousse, or other flammable hair products. Tie back long hair and tie down loose clothing.
3. Do NOT wear sandals or other open-toed shoes in the lab.
4. Remove jewelry on hands and wrists. Loose jewelry, such as chains and long necklaces, should be removed to prevent them from getting caught in equipment.
5. Do not taste any substances or draw any material into a tube with your mouth.
6. Proper behavior is expected in the lab. Practical jokes and fooling around can lead to accidents and injury.
7. Keep your work area uncluttered.

Laboratory Work

1. Collect and carry all equipment and materials to your work area before beginning a lab.
2. Remain in your own work area unless given permission by your teacher to leave it.

SR-12 • Science Skill Handbook

3. Always slant test tubes away from yourself and others when heating them, adding substances to them, or rinsing them.
4. If instructed to smell a substance in a container, hold the container a short distance away and fan vapors toward your nose.
5. Do NOT substitute other chemicals/substances for those in the materials list unless instructed to do so by your teacher.
6. Do NOT take any materials or chemicals outside of the laboratory.
7. Stay out of storage areas unless instructed to be there and supervised by your teacher.

Laboratory Cleanup

1. Turn off all burners, water, and gas, and disconnect all electrical devices.
2. Clean all pieces of equipment and return all materials to their proper places.
3. Dispose of chemicals and other materials as directed by your teacher. Place broken glass and solid substances in the proper containers. Never discard materials in the sink.
4. Clean your work area.
5. Wash your hands with soap and water thoroughly BEFORE removing your goggles.

Emergencies

1. Report any fire, electrical shock, glassware breakage, spill, or injury, no matter how small, to your teacher immediately. Follow his or her instructions.
2. If your clothing should catch fire, STOP, DROP, and ROLL. If possible, smother it with the fire blanket or get under a safety shower. NEVER RUN.
3. If a fire should occur, turn off all gas and leave the room according to established procedures.
4. In most instances, your teacher will clean up spills. Do NOT attempt to clean up spills unless you are given permission and instructions to do so.
5. If chemicals come into contact with your eyes or skin, notify your teacher immediately. Use the eyewash, or flush your skin or eyes with large quantities of water.
6. The fire extinguisher and first-aid kit should only be used by your teacher unless it is an extreme emergency and you have been given permission.
7. If someone is injured or becomes ill, only a professional medical provider or someone certified in first aid should perform first-aid procedures.

Science Skill Handbook • **SR-13**

Math Skill Handbook

Math Review

Use Fractions

A fraction compares a part to a whole. In the fraction $\frac{2}{3}$, the 2 represents the part and is the numerator. The 3 represents the whole and is the denominator.

Reduce Fractions To reduce a fraction, you must find the largest factor that is common to both the numerator and the denominator, the greatest common factor (GCF). Divide both numbers by the GCF. The fraction has then been reduced, or it is in its simplest form.

Example

Twelve of the 20 chemicals in the science lab are in powder form. What fraction of the chemicals used in the lab are in powder form?

Step 1 Write the fraction.

$$\frac{\text{part}}{\text{whole}} = \frac{12}{20}$$

Step 2 To find the GCF of the numerator and denominator, list all of the factors of each number.

Factors of 12: 1, 2, 3, 4, 6, 12 (the numbers that divide evenly into 12)

Factors of 20: 1, 2, 4, 5, 10, 20 (the numbers that divide evenly into 20)

Step 3 List the common factors.

1, 2, 4

Step 4 Choose the greatest factor in the list. The GCF of 12 and 20 is 4.

Step 5 Divide the numerator and denominator by the GCF.

$$\frac{12 \div 4}{20 \div 4} = \frac{3}{5}$$

In the lab, $\frac{3}{5}$ of the chemicals are in powder form.

Practice Problem At an amusement park, 66 of 90 rides have a height restriction. What fraction of the rides, in its simplest form, has a height restriction?

Add and Subtract Fractions with Like Denominators
To add or subtract fractions with the same denominator, add or subtract the numerators and write the sum or difference over the denominator. After finding the sum or difference, find the simplest form for your fraction.

Example 1

In the forest outside your house, $\frac{1}{8}$ of the animals are rabbits, $\frac{3}{8}$ are squirrels, and the remainder are birds and insects. How many are mammals?

Step 1 Add the numerators.

$$\frac{1}{8} + \frac{3}{8} = \frac{(1+3)}{8} = \frac{4}{8}$$

Step 2 Find the GCF.

$$\frac{4}{8} \text{ (GCF, 4)}$$

Step 3 Divide the numerator and denominator by the GCF.

$$\frac{4 \div 4}{8 \div 4} = \frac{1}{2}$$

$\frac{1}{2}$ of the animals are mammals.

Example 2

If $\frac{7}{16}$ of the Earth is covered by freshwater, and $\frac{1}{16}$ of that is in glaciers, how much freshwater is not frozen?

Step 1 Subtract the numerators.

$$\frac{7}{16} - \frac{1}{16} = \frac{(7-1)}{16} = \frac{6}{16}$$

Step 2 Find the GCF.

$$\frac{6}{16} \text{ (GCF, 2)}$$

Step 3 Divide the numerator and denominator by the GCF.

$$\frac{6 \div 2}{16 \div 2} = \frac{3}{8}$$

$\frac{3}{8}$ of the freshwater is not frozen.

Practice Problem A bicycle rider is riding at a rate of 15 km/h for $\frac{4}{9}$ of his ride, 10 km/h for $\frac{2}{9}$ of his ride, and 8 km/h for the remainder of the ride. How much of his ride is he riding at a rate greater than 8 km/h?

Add and Subtract Fractions with Unlike Denominators To add or subtract fractions with unlike denominators, first find the least common denominator (LCD). This is the smallest number that is a common multiple of both denominators. Rename each fraction with the LCD, and then add or subtract. Find the simplest form if necessary.

Example 1

A chemist makes a paste that is $\frac{1}{2}$ table salt (NaCl), $\frac{1}{3}$ sugar ($C_6H_{12}O_6$), and the remainder is water (H_2O). How much of the paste is a solid?

Step 1 Find the LCD of the fractions.

$\frac{1}{2} + \frac{1}{3}$ (LCD, 6)

Step 2 Rename each numerator and each denominator with the LCD.

Step 3 Add the numerators.

$\frac{3}{6} + \frac{2}{6} = \frac{(3+2)}{6} = \frac{5}{6}$

$\frac{5}{6}$ of the paste is a solid.

Example 2

The average precipitation in Grand Junction, CO, is $\frac{7}{10}$ inch in November, and $\frac{3}{5}$ inch in December. What is the total average precipitation?

Step 1 Find the LCD of the fractions.

$\frac{7}{10} + \frac{3}{5}$ (LCD, 10)

Step 2 Rename each numerator and each denominator with the LCD.

Step 3 Add the numerators.

$\frac{7}{10} + \frac{6}{10} = \frac{(7+6)}{10} = \frac{13}{10}$

$\frac{13}{10}$ inches total precipitation, or $1\frac{3}{10}$ inches.

Practice Problem On an electric bill, about $\frac{1}{8}$ of the energy is from solar energy and about $\frac{1}{10}$ is from wind power. How much of the total bill is from solar energy and wind power combined?

Example 3

In your body, $\frac{7}{10}$ of your muscle contractions are involuntary (cardiac and smooth muscle tissue). Smooth muscle makes $\frac{3}{15}$ of your muscle contractions. How many of your muscle contractions are made by cardiac muscle?

Step 1 Find the LCD of the fractions.

$\frac{7}{10} - \frac{3}{15}$ (LCD, 30)

Step 2 Rename each numerator and each denominator with the LCD.

$\frac{7 \times 3}{10 \times 3} = \frac{21}{30}$

$\frac{3 \times 2}{15 \times 2} = \frac{6}{30}$

Step 3 Subtract the numerators.

$\frac{21}{30} - \frac{6}{30} = \frac{(21-6)}{30} = \frac{15}{30}$

Step 4 Find the GCF.

$\frac{15}{30}$ (GCF, 15)

$\frac{1}{2}$

$\frac{1}{2}$ of all muscle contractions are cardiac muscle.

Example 4

Tony wants to make cookies that call for $\frac{3}{4}$ of a cup of flour, but he only has $\frac{1}{3}$ of a cup. How much more flour does he need?

Step 1 Find the LCD of the fractions.

$\frac{3}{4} - \frac{1}{3}$ (LCD, 12)

Step 2 Rename each numerator and each denominator with the LCD.

$\frac{3 \times 3}{4 \times 3} = \frac{9}{12}$

$\frac{1 \times 4}{3 \times 4} = \frac{4}{12}$

Step 3 Subtract the numerators.

$\frac{9}{12} - \frac{4}{12} = \frac{(9-4)}{12} = \frac{5}{12}$

$\frac{5}{12}$ of a cup of flour

Practice Problem Using the information provided to you in Example 3 above, determine how many muscle contractions are voluntary (skeletal muscle).

Math Skill Handbook • SR-15

Multiply Fractions To multiply with fractions, multiply the numerators and multiply the denominators. Find the simplest form if necessary.

> **Example**
>
> Multiply $\frac{3}{5}$ by $\frac{1}{3}$.
>
> **Step 1** Multiply the numerators and denominators.
>
> $$\frac{3}{5} \times \frac{1}{3} = \frac{(3 \times 1)}{(5 \times 3)} \frac{3}{15}$$
>
> **Step 2** Find the GCF.
>
> $\frac{3}{15}$ (GCF, 3)
>
> **Step 3** Divide the numerator and denominator by the GCF.
>
> $$\frac{3 \div 3}{15 \div 3} = \frac{1}{5}$$
>
> $\frac{3}{5}$ multiplied by $\frac{1}{3}$ is $\frac{1}{5}$.

Practice Problem Multiply $\frac{3}{14}$ by $\frac{5}{16}$.

Find a Reciprocal Two numbers whose product is 1 are called multiplicative inverses, or reciprocals.

> **Example**
>
> Find the reciprocal of $\frac{3}{8}$.
>
> **Step 1** Inverse the fraction by putting the denominator on top and the numerator on the bottom.
>
> $\frac{8}{3}$
>
> The reciprocal of $\frac{3}{8}$ is $\frac{8}{3}$.

Practice Problem Find the reciprocal of $\frac{4}{9}$.

Divide Fractions To divide one fraction by another fraction, multiply the dividend by the reciprocal of the divisor. Find the simplest form if necessary.

> **Example 1**
>
> Divide $\frac{1}{9}$ by $\frac{1}{3}$.
>
> **Step 1** Find the reciprocal of the divisor.
>
> The reciprocal of $\frac{1}{3}$ is $\frac{3}{1}$.
>
> **Step 2** Multiply the dividend by the reciprocal of the divisor.
>
> $$\frac{\frac{1}{9}}{\frac{1}{3}} = \frac{1}{9} \times \frac{3}{1} = \frac{(1 \times 3)}{(9 \times 1)} = \frac{3}{9}$$
>
> **Step 3** Find the GCF.
>
> $\frac{3}{9}$ (GCF, 3)
>
> **Step 4** Divide the numerator and denominator by the GCF.
>
> $$\frac{3 \div 3}{9 \div 3} = \frac{1}{3}$$
>
> $\frac{1}{9}$ divided by $\frac{1}{3}$ is $\frac{1}{3}$.

> **Example 2**
>
> Divide $\frac{3}{5}$ by $\frac{1}{4}$.
>
> **Step 1** Find the reciprocal of the divisor.
>
> The reciprocal of $\frac{1}{4}$ is $\frac{4}{1}$.
>
> **Step 2** Multiply the dividend by the reciprocal of the divisor.
>
> $$\frac{\frac{3}{5}}{\frac{1}{4}} = \frac{3}{5} \times \frac{4}{1} = \frac{(3 \times 4)}{(5 \times 1)} = \frac{12}{5}$$
>
> $\frac{3}{5}$ divided by $\frac{1}{4}$ is $\frac{12}{5}$ or $2\frac{2}{5}$.

Practice Problem Divide $\frac{3}{11}$ by $\frac{7}{10}$.

Use Ratios

When you compare two numbers by division, you are using a ratio. Ratios can be written 3 to 5, 3:5, or $\frac{3}{5}$. Ratios, like fractions, also can be written in simplest form.

Ratios can represent one type of probability, called odds. This is a ratio that compares the number of ways a certain outcome occurs to the number of possible outcomes. For example, if you flip a coin 100 times, what are the odds that it will come up heads? There are two possible outcomes, heads or tails, so the odds of coming up heads are 50:100. Another way to say this is that 50 out of 100 times the coin will come up heads. In its simplest form, the ratio is 1:2.

Example 1

A chemical solution contains 40 g of salt and 64 g of baking soda. What is the ratio of salt to baking soda as a fraction in simplest form?

Step 1 Write the ratio as a fraction.
$$\frac{salt}{baking\ soda} = \frac{40}{64}$$

Step 2 Express the fraction in simplest form. The GCF of 40 and 64 is 8.
$$\frac{40}{64} = \frac{40 \div 8}{64 \div 8} = \frac{5}{8}$$

The ratio of salt to baking soda in the sample is 5:8.

Example 2

Sean rolls a 6-sided die 6 times. What are the odds that the side with a 3 will show?

Step 1 Write the ratio as a fraction.
$$\frac{number\ of\ sides\ with\ a\ 3}{number\ of\ possible\ sides} = \frac{1}{6}$$

Step 2 Multiply by the number of attempts.
$$\frac{1}{6} \times 6\ attempts = \frac{6}{6}\ attempts = 1\ attempt$$

1 attempt out of 6 will show a 3.

Practice Problem Two metal rods measure 100 cm and 144 cm in length. What is the ratio of their lengths in simplest form?

Use Decimals

A fraction with a denominator that is a power of ten can be written as a decimal. For example, 0.27 means $\frac{27}{100}$. The decimal point separates the ones place from the tenths place.

Any fraction can be written as a decimal using division. For example, the fraction $\frac{5}{8}$ can be written as a decimal by dividing 5 by 8. Written as a decimal, it is 0.625.

Add or Subtract Decimals When adding and subtracting decimals, line up the decimal points before carrying out the operation.

Example 1

Find the sum of 47.68 and 7.80.

Step 1 Line up the decimal places when you write the numbers.

```
  47.68
+  7.80
```

Step 2 Add the decimals.

```
  1 1
  47.68
+  7.80
  55.48
```

The sum of 47.68 and 7.80 is 55.48.

Example 2

Find the difference of 42.17 and 15.85.

Step 1 Line up the decimal places when you write the number.

```
  42.17
- 15.85
```

Step 2 Subtract the decimals.

```
  3 11 1
  42.17
- 15.85
  26.32
```

The difference of 42.17 and 15.85 is 26.32.

Practice Problem Find the sum of 1.245 and 3.842.

Math Skill Handbook • SR-17

Multiply Decimals To multiply decimals, multiply the numbers like numbers without decimal points. Count the decimal places in each factor. The product will have the same number of decimal places as the sum of the decimal places in the factors.

> **Example**
>
> Multiply 2.4 by 5.9.
>
> **Step 1** Multiply the factors like two whole numbers.
>
> $24 \times 59 = 1416$
>
> **Step 2** Find the sum of the number of decimal places in the factors. Each factor has one decimal place, for a sum of two decimal places.
>
> **Step 3** The product will have two decimal places.
>
> 14.16
>
> The product of 2.4 and 5.9 is 14.16.

Practice Problem Multiply 4.6 by 2.2.

Divide Decimals When dividing decimals, change the divisor to a whole number. To do this, multiply both the divisor and the dividend by the same power of ten. Then place the decimal point in the quotient directly above the decimal point in the dividend. Then divide as you do with whole numbers.

> **Example**
>
> Divide 8.84 by 3.4.
>
> **Step 1** Multiply both factors by 10.
>
> $3.4 \times 10 = 34$, $8.84 \times 10 = 88.4$
>
> **Step 2** Divide 88.4 by 34.
>
> ```
>       2.6
>   34)88.4
>     -68
>      204
>     -204
>        0
> ```
>
> 8.84 divided by 3.4 is 2.6.

Practice Problem Divide 75.6 by 3.6.

Use Proportions

An equation that shows that two ratios are equivalent is a proportion. The ratios $\frac{2}{4}$ and $\frac{5}{10}$ are equivalent, so they can be written as $\frac{2}{4} = \frac{5}{10}$. This equation is a proportion.

When two ratios form a proportion, the cross products are equal. To find the cross products in the proportion $\frac{2}{4} = \frac{5}{10}$, multiply the 2 and the 10, and the 4 and the 5. Therefore $2 \times 10 = 4 \times 5$, or $20 = 20$.

Because you know that both ratios are equal, you can use cross products to find a missing term in a proportion. This is known as solving the proportion.

> **Example**
>
> The heights of a tree and a pole are proportional to the lengths of their shadows. The tree casts a shadow of 24 m when a 6-m pole casts a shadow of 4 m. What is the height of the tree?
>
> **Step 1** Write a proportion.
>
> $\frac{\text{height of tree}}{\text{height of pole}} = \frac{\text{length of tree's shadow}}{\text{length of pole's shadow}}$
>
> **Step 2** Substitute the known values into the proportion. Let h represent the unknown value, the height of the tree.
>
> $\frac{h}{6} \times \frac{24}{4}$
>
> **Step 3** Find the cross products.
>
> $h \times 4 = 6 \times 24$
>
> **Step 4** Simplify the equation.
>
> $4h \times 144$
>
> **Step 5** Divide each side by 4.
>
> $\frac{4h}{4} \times \frac{144}{4}$
>
> $h = 36$
>
> The height of the tree is 36 m.

Practice Problem The ratios of the weights of two objects on the Moon and on Earth are in proportion. A rock weighing 3 N on the Moon weighs 18 N on Earth. How much would a rock that weighs 5 N on the Moon weigh on Earth?

Use Percentages

The word *percent* means "out of one hundred." It is a ratio that compares a number to 100. Suppose you read that 77 percent of Earth's surface is covered by water. That is the same as reading that the fraction of Earth's surface covered by water is $\frac{77}{100}$. To express a fraction as a percent, first find the equivalent decimal for the fraction. Then, multiply the decimal by 100 and add the percent symbol.

Example 1

Express $\frac{13}{20}$ as a percent.

Step 1 Find the equivalent decimal for the fraction.

$$\begin{array}{r} 0.65 \\ 20\overline{)13.00} \\ \underline{12\ 0} \\ 1\ 00 \\ \underline{1\ 00} \\ 0 \end{array}$$

Step 2 Rewrite the fraction $\frac{13}{20}$ as 0.65.

Step 3 Multiply 0.65 by 100 and add the % symbol.

$$0.65 \times 100 = 65 = 65\%$$

So, $\frac{13}{20} = 65\%$.

This also can be solved as a proportion.

Example 2

Express $\frac{13}{20}$ as a percent.

Step 1 Write a proportion.

$$\frac{13}{20} = \frac{x}{100}$$

Step 2 Find the cross products.

$$1300 = 20x$$

Step 3 Divide each side by 20.

$$\frac{1300}{20} = \frac{20x}{20}$$

$$65\% = x$$

Practice Problem In one year, 73 of 365 days were rainy in one city. What percent of the days in that city were rainy?

Solve One-Step Equations

A statement that two expressions are equal is an equation. For example, $A = B$ is an equation that states that A is equal to B.

An equation is solved when a variable is replaced with a value that makes both sides of the equation equal. To make both sides equal the inverse operation is used. Addition and subtraction are inverses, and multiplication and division are inverses.

Example 1

Solve the equation $x - 10 = 35$.

Step 1 Find the solution by adding 10 to each side of the equation.

$$x - 10 = 35$$
$$x - 10 + 10 = 35 - 10$$
$$x = 45$$

Step 2 Check the solution.

$$x - 10 = 35$$
$$45 - 10 = 35$$
$$35 = 35$$

Both sides of the equation are equal, so $x = 45$.

Example 2

In the formula $a = bc$, find the value of c if $a = 20$ and $b = 2$.

Step 1 Rearrange the formula so the unknown value is by itself on one side of the equation by dividing both sides by b.

$$a = bc$$
$$\frac{a}{b} = \frac{bc}{b}$$
$$\frac{a}{b} = c$$

Step 2 Replace the variables a and b with the values that are given.

$$\frac{a}{b} = c$$
$$\frac{20}{2} = c$$
$$10 = c$$

Step 3 Check the solution.

$$a = bc$$
$$20 = 2 \times 10$$
$$20 = 20$$

Both sides of the equation are equal, so $c = 10$ is the solution when $a = 20$ and $b = 2$.

Practice Problem In the formula $h = gd$, find the value of d if $g = 12.3$ and $h = 17.4$.

Use Statistics

The branch of mathematics that deals with collecting, analyzing, and presenting data is statistics. In statistics, there are three common ways to summarize data with a single number—the mean, the median, and the mode.

The **mean** of a set of data is the arithmetic average. It is found by adding the numbers in the data set and dividing by the number of items in the set.

The **median** is the middle number in a set of data when the data are arranged in numerical order. If there were an even number of data points, the median would be the mean of the two middle numbers.

The **mode** of a set of data is the number or item that appears most often.

Another number that often is used to describe a set of data is the range. The **range** is the difference between the largest number and the smallest number in a set of data.

Example

The speeds (in m/s) for a race car during five different time trials are 39, 37, 44, 36, and 44.

To find the mean:

Step 1   Find the sum of the numbers.

$39 + 37 + 44 + 36 + 44 = 200$

Step 2   Divide the sum by the number of items, which is 5.

$200 \div 5 = 40$

The mean is 40 m/s.

To find the median:

Step 1   Arrange the measures from least to greatest.

36, 37, 39, 44, 44

Step 2   Determine the middle measure.

36, 37, <u>39</u>, 44, 44

The median is 39 m/s.

To find the mode:

Step 1   Group the numbers that are the same together.

44, 44, 36, 37, 39

Step 2   Determine the number that occurs most in the set.

<u>44, 44</u>, 36, 37, 39

The mode is 44 m/s.

To find the range:

Step 1   Arrange the measures from greatest to least.

44, 44, 39, 37, 36

Step 2   Determine the greatest and least measures in the set.

<u>44</u>, 44, 39, 37, <u>36</u>

Step 3   Find the difference between the greatest and least measures.

$44 - 36 = 8$

The range is 8 m/s.

Practice Problem Find the mean, median, mode, and range for the data set 8, 4, 12, 8, 11, 14, 16.

A **frequency table** shows how many times each piece of data occurs, usually in a survey. **Table 1** below shows the results of a student survey on favorite color.

Table 1 Student Color Choice		
Color	Tally	Frequency
red	IIII	4
blue	IIII	5
black	II	2
green	III	3
purple	IIII II	7
yellow	IIII I	6

Based on the frequency table data, which color is the favorite?

SR-20 • Math Skill Handbook

Use Geometry

The branch of mathematics that deals with the measurement, properties, and relationships of points, lines, angles, surfaces, and solids is called geometry.

Perimeter The **perimeter** (P) is the distance around a geometric figure. To find the perimeter of a rectangle, add the length and width and multiply that sum by two, or $2(l + w)$. To find perimeters of irregular figures, add the length of the sides.

Example 1

Find the perimeter of a rectangle that is 3 m long and 5 m wide.

Step 1 You know that the perimeter is 2 times the sum of the width and length.

$P = 2(3 \text{ m} + 5 \text{ m})$

Step 2 Find the sum of the width and length.

$P = 2(8 \text{ m})$

Step 3 Multiply by 2.

$P = 16 \text{ m}$

The perimeter is 16 m.

Example 2

Find the perimeter of a shape with sides measuring 2 cm, 5 cm, 6 cm, 3 cm.

Step 1 You know that the perimeter is the sum of all the sides.

$P = 2 + 5 + 6 + 3$

Step 2 Find the sum of the sides.

$P = 2 + 5 + 6 + 3$

$P = 16$

The perimeter is 16 cm.

Practice Problem Find the perimeter of a rectangle with a length of 18 m and a width of 7 m.

Practice Problem Find the perimeter of a triangle measuring 1.6 cm by 2.4 cm by 2.4 cm.

Area of a Rectangle The **area** (A) is the number of square units needed to cover a surface. To find the area of a rectangle, multiply the length times the width, or $l \times w$. When finding area, the units also are multiplied. Area is given in square units.

Example

Find the area of a rectangle with a length of 1 cm and a width of 10 cm.

Step 1 You know that the area is the length multiplied by the width.

$A = (1 \text{ cm} \times 10 \text{ cm})$

Step 2 Multiply the length by the width. Also multiply the units.

$A = 10 \text{ cm}^2$

The area is 10 cm².

Practice Problem Find the area of a square whose sides measure 4 m.

Area of a Triangle To find the area of a triangle, use the formula:

$A = \frac{1}{2}(\text{base} \times \text{height})$

The base of a triangle can be any of its sides. The height is the perpendicular distance from a base to the opposite endpoint, or vertex.

Example

Find the area of a triangle with a base of 18 m and a height of 7 m.

Step 1 You know that the area is $\frac{1}{2}$ the base times the height.

$A = \frac{1}{2}(18 \text{ m} \times 7 \text{ m})$

Step 2 Multiply $\frac{1}{2}$ by the product of 18×7. Multiply the units.

$A = \frac{1}{2}(126 \text{ m}^2)$

$A = 63 \text{ m}^2$

The area is 63 m².

Practice Problem Find the area of a triangle with a base of 27 cm and a height of 17 cm.

Circumference of a Circle The **diameter** (d) of a circle is the distance across the circle through its center, and the **radius** (r) is the distance from the center to any point on the circle. The radius is half of the diameter. The distance around the circle is called the **circumference** (C). The formula for finding the circumference is:

$C = 2\pi r$ or $C = \pi d$

The circumference divided by the diameter is always equal to 3.1415926… This nonterminating and nonrepeating number is represented by the Greek letter π (pi). An approximation often used for π is 3.14.

Example 1

Find the circumference of a circle with a radius of 3 m.

Step 1 You know the formula for the circumference is 2 times the radius times π.

$C = 2\pi(3)$

Step 2 Multiply 2 times the radius.

$C = 6\pi$

Step 3 Multiply by π.

$C \approx 19$ m

The circumference is about 19 m.

Example 2

Find the circumference of a circle with a diameter of 24.0 cm.

Step 1 You know the formula for the circumference is the diameter times π.

$C = \pi(24.0)$

Step 2 Multiply the diameter by π.

$C \approx 75.4$ cm

The circumference is about 75.4 cm.

Practice Problem Find the circumference of a circle with a radius of 19 cm.

Area of a Circle The formula for the area of a circle is: $A = \pi r^2$

Example 1

Find the area of a circle with a radius of 4.0 cm.

Step 1 $A = \pi(4.0)^2$

Step 2 Find the square of the radius.

$A = 16\pi$

Step 3 Multiply the square of the radius by π.

$A \approx 50$ cm^2

The area of the circle is about 50 cm^2.

Example 2

Find the area of a circle with a radius of 225 m.

Step 1 $A = \pi(225)^2$

Step 2 Find the square of the radius.

$A = 50625\pi$

Step 3 Multiply the square of the radius by π.

$A \approx 159043.1$

The area of the circle is about 159043.1 m^2.

Example 3

Find the area of a circle whose diameter is 20.0 mm.

Step 1 Remember that the radius is half of the diameter.

$A = \pi\left(\frac{20.0}{2}\right)^2$

Step 2 Find the radius.

$A = \pi(10.0)^2$

Step 3 Find the square of the radius.

$A = 100\pi$

Step 4 Multiply the square of the radius by π.

$A \approx 314$ mm^2

The area of the circle is about 314 mm^2.

Practice Problem Find the area of a circle with a radius of 16 m.

Volume The measure of space occupied by a solid is the **volume** (V). To find the volume of a rectangular solid multiply the length times width times height, or $V = l \times w \times h$. It is measured in cubic units, such as cubic centimeters (cm^3).

Example

Find the volume of a rectangular solid with a length of 2.0 m, a width of 4.0 m, and a height of 3.0 m.

Step 1 You know the formula for volume is the length times the width times the height.

$V = 2.0 \text{ m} \times 4.0 \text{ m} \times 3.0 \text{ m}$

Step 2 Multiply the length times the width times the height.

$V = 24 \text{ m}^3$

The volume is 24 m^3.

Practice Problem Find the volume of a rectangular solid that is 8 m long, 4 m wide, and 4 m high.

To find the volume of other solids, multiply the area of the base times the height.

Example 1

Find the volume of a solid that has a triangular base with a length of 8.0 m and a height of 7.0 m. The height of the entire solid is 15.0 m.

Step 1 You know that the base is a triangle, and the area of a triangle is $\frac{1}{2}$ the base times the height, and the volume is the area of the base times the height.

$V = \left[\frac{1}{2}(b \times h)\right] \times 15$

Step 2 Find the area of the base.

$V = \left[\frac{1}{2}(8 \times 7)\right] \times 15$

$V = \left(\frac{1}{2} \times 56\right) \times 15$

Step 3 Multiply the area of the base by the height of the solid.

$V = 28 \times 15$

$V = 420 \text{ m}^3$

The volume is 420 m^3.

Example 2

Find the volume of a cylinder that has a base with a radius of 12.0 cm, and a height of 21.0 cm.

Step 1 You know that the base is a circle, and the area of a circle is the square of the radius times π, and the volume is the area of the base times the height.

$V = (\pi r^2) \times 21$

$V = (\pi 12^2) \times 21$

Step 2 Find the area of the base.

$V = 144\pi \times 21$

$V = 452 \times 21$

Step 3 Multiply the area of the base by the height of the solid.

$V \approx 9{,}500 \text{ cm}^3$

The volume is about 9,500 cm^3.

Example 3

Find the volume of a cylinder that has a diameter of 15 mm and a height of 4.8 mm.

Step 1 You know that the base is a circle with an area equal to the square of the radius times π. The radius is one-half the diameter. The volume is the area of the base times the height.

$V = (\pi r^2) \times 4.8$

$V = \left[\pi\left(\frac{1}{2} \times 15\right)^2\right] \times 4.8$

$V = (\pi 7.5^2) \times 4.8$

Step 2 Find the area of the base.

$V = 56.25\pi \times 4.8$

$V \approx 176.71 \times 4.8$

Step 3 Multiply the area of the base by the height of the solid.

$V \approx 848.2$

The volume is about 848.2 mm^3.

Practice Problem Find the volume of a cylinder with a diameter of 7 cm in the base and a height of 16 cm.

Science Applications

Measure in SI

The metric system of measurement was developed in 1795. A modern form of the metric system, called the International System (SI), was adopted in 1960 and provides the standard measurements that all scientists around the world can understand.

The SI system is convenient because unit sizes vary by powers of 10. Prefixes are used to name units. Look at **Table 2** for some common SI prefixes and their meanings.

Table 2 Common SI Prefixes

Prefix	Symbol	Meaning	
kilo–	k	1,000	thousandth
hecto–	h	100	hundred
deka–	da	10	ten
deci–	d	0.1	tenth
centi–	c	0.01	hundreth
milli–	m	0.001	thousandth

Example

How many grams equal one kilogram?

Step 1 Find the prefix *kilo–* in **Table 2**.

Step 2 Using **Table 2,** determine the meaning of *kilo–*. According to the table, it means 1,000. When the prefix *kilo–* is added to a unit, it means that there are 1,000 of the units in a "kilounit."

Step 3 Apply the prefix to the units in the question. The units in the question are grams. There are 1,000 grams in a kilogram.

Practice Problem Is a milligram larger or smaller than a gram? How many of the smaller units equal one larger unit? What fraction of the larger unit does one smaller unit represent?

Dimensional Analysis

Convert SI Units In science, quantities such as length, mass, and time sometimes are measured using different units. A process called dimensional analysis can be used to change one unit of measure to another. This process involves multiplying your starting quantity and units by one or more conversion factors. A conversion factor is a ratio equal to one and can be made from any two equal quantities with different units. If 1,000 mL equal 1 L then two ratios can be made.

$$\frac{1,000 \text{ mL}}{1 \text{ L}} = \frac{1 \text{ L}}{1,000 \text{ mL}} = 1$$

One can convert between units in the SI system by using the equivalents in **Table 2** to make conversion factors.

Example

How many cm are in 4 m?

Step 1 Write conversion factors for the units given. From **Table 2,** you know that 100 cm = 1 m. The conversion factors are

$$\frac{100 \text{ cm}}{1 \text{ m}} \text{ and } \frac{1 \text{ m}}{100 \text{ cm}}$$

Step 2 Decide which conversion factor to use. Select the factor that has the units you are converting from (m) in the denominator and the units you are converting to (cm) in the numerator.

$$\frac{100 \text{ cm}}{1 \text{ m}}$$

Step 3 Multiply the starting quantity and units by the conversion factor. Cancel the starting units with the units in the denominator. There are 400 cm in 4 m.

$$4 \text{ m} \times \frac{100 \text{ cm}}{1 \text{ m}} = 400 \text{ cm}$$

Practice Problem How many milligrams are in one kilogram? (Hint: You will need to use two conversion factors from **Table 2**.)

Table 3 Unit System Equivalents

Type of Measurement	Equivalent
Length	1 in = 2.54 cm 1 yd = 0.91 m 1 mi = 1.61 km
Mass and weight*	1 oz = 28.35 g 1 lb = 0.45 kg 1 ton (short) = 0.91 tonnes (metric tons) 1 lb = 4.45 N
Volume	1 in^3 = 16.39 cm^3 1 qt = 0.95 L 1 gal = 3.78 L
Area	1 in^2 = 6.45 cm^2 1 yd^2 = 0.83 m^2 1 mi^2 = 2.59 km^2 1 acre = 0.40 hectares
Temperature	°C = $\frac{(°F - 32)}{1.8}$ K = °C + 273

*Weight is measured in standard Earth gravity.

Convert Between Unit Systems Table 3 gives a list of equivalents that can be used to convert between English and SI units.

Example

If a meterstick has a length of 100 cm, how long is the meterstick in inches?

Step 1 Write the conversion factors for the units given. From **Table 3,** 1 in = 2.54 cm.

$$\frac{1 \text{ in}}{2.54 \text{ cm}} \text{ and } \frac{2.54 \text{ cm}}{1 \text{ in}}$$

Step 2 Determine which conversion factor to use. You are converting from cm to in. Use the conversion factor with cm on the bottom.

$$\frac{1 \text{ in}}{2.54 \text{ cm}}$$

Step 3 Multiply the starting quantity and units by the conversion factor. Cancel the starting units with the units in the denominator. Round your answer to the nearest tenth.

$$100 \text{ cm} \times \frac{1 \text{ in}}{2.54 \text{ cm}} = 39.37 \text{ in}$$

The meterstick is about 39.4 in long.

Practice Problem 1 A book has a mass of 5 lb. What is the mass of the book in kg?

Practice Problem 2 Use the equivalent for in and cm (1 in = 2.54 cm) to show how 1 in^3 ≈ 16.39 cm^3.

Precision and Significant Digits

When you make a measurement, the value you record depends on the precision of the measuring instrument. This precision is represented by the number of significant digits recorded in the measurement. When counting the number of significant digits, all digits are counted except zeros at the end of a number with no decimal point such as 2,050, and zeros at the beginning of a decimal such as 0.03020. When adding or subtracting numbers with different precision, round the answer to the smallest number of decimal places of any number in the sum or difference. When multiplying or dividing, the answer is rounded to the smallest number of significant digits of any number being multiplied or divided.

Example

The lengths 5.28 and 5.2 are measured in meters. Find the sum of these lengths and record your answer using the correct number of significant digits.

Step 1 Find the sum.

  5.28 m   2 digits after the decimal
+ 5.2 m   1 digit after the decimal
  10.48 m

Step 2 Round to one digit after the decimal because the least number of digits after the decimal of the numbers being added is 1.

The sum is 10.5 m.

Practice Problem 1 How many significant digits are in the measurement 7,071,301 m? How many significant digits are in the measurement 0.003010 g?

Practice Problem 2 Multiply 5.28 and 5.2 using the rule for multiplying and dividing. Record the answer using the correct number of significant digits.

Scientific Notation

Many times numbers used in science are very small or very large. Because these numbers are difficult to work with scientists use scientific notation. To write numbers in scientific notation, move the decimal point until only one non-zero digit remains on the left. Then count the number of places you moved the decimal point and use that number as a power of ten. For example, the average distance from the Sun to Mars is 227,800,000,000 m. In scientific notation, this distance is 2.278×10^{11} m. Because you moved the decimal point to the left, the number is a positive power of ten.

The mass of an electron is about 0.000 000 000 000 000 000 000 000 000 000 911 kg. Expressed in scientific notation, this mass is 9.11×10^{-31} kg. Because the decimal point was moved to the right, the number is a negative power of ten.

Example

Earth is 149,600,000 km from the Sun. Express this in scientific notation.

Step 1 Move the decimal point until one non-zero digit remains on the left.

1.496 000 00

Step 2 Count the number of decimal places you have moved. In this case, eight.

Step 2 Show that number as a power of ten, 10^8.

Earth is 1.496×10^8 km from the Sun.

Practice Problem 1 How many significant digits are in 149,600,000 km? How many significant digits are in 1.496×10^8 km?

Practice Problem 2 Parts used in a high performance car must be measured to 7×10^{-6} m. Express this number as a decimal.

Practice Problem 3 A CD is spinning at 539 revolutions per minute. Express this number in scientific notation.

Make and Use Graphs

Data in tables can be displayed in a graph—a visual representation of data. Common graph types include line graphs, bar graphs, and circle graphs.

Line Graph A line graph shows a relationship between two variables that change continuously. The independent variable is changed and is plotted on the x-axis. The dependent variable is observed, and is plotted on the y-axis.

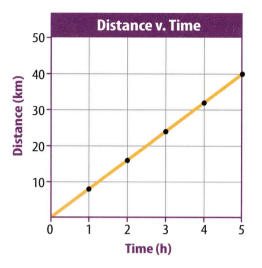

Example

Draw a line graph of the data below from a cyclist in a long-distance race.

Table 4 Bicycle Race Data	
Time (h)	Distance (km)
0	0
1	8
2	16
3	24
4	32
5	40

Step 1 Determine the x-axis and y-axis variables. Time varies independently of distance and is plotted on the x-axis. Distance is dependent on time and is plotted on the y-axis.

Step 2 Determine the scale of each axis. The x-axis data ranges from 0 to 5. The y-axis data ranges from 0 to 50.

Step 3 Using graph paper, draw and label the axes. Include units in the labels.

Step 4 Draw a point at the intersection of the time value on the x-axis and corresponding distance value on the y-axis. Connect the points and label the graph with a title, as shown in **Figure 8**.

Figure 8 This line graph shows the relationship between distance and time during a bicycle ride.

Practice Problem A puppy's shoulder height is measured during the first year of her life. The following measurements were collected: (3 mo, 52 cm), (6 mo, 72 cm), (9 mo, 83 cm), (12 mo, 86 cm). Graph this data.

Find a Slope The slope of a straight line is the ratio of the vertical change, rise, to the horizontal change, run.

$$\text{Slope} = \frac{\text{vertical change (rise)}}{\text{horizontal change (run)}} = \frac{\text{change in } y}{\text{change in } x}$$

Example

Find the slope of the graph in **Figure 8**.

Step 1 You know that the slope is the change in y divided by the change in x.

$$\text{Slope} = \frac{\text{change in } y}{\text{change in } x}$$

Step 2 Determine the data points you will be using. For a straight line, choose the two sets of points that are the farthest apart.

$$\text{Slope} = \frac{(40 - 0) \text{ km}}{(5 - 0) \text{ h}}$$

Step 3 Find the change in y and x.

$$\text{Slope} = \frac{40 \text{ km}}{5 \text{ h}}$$

Step 4 Divide the change in y by the change in x.

$$\text{Slope} = \frac{8 \text{ km}}{\text{h}}$$

The slope of the graph is 8 km/h.

Bar Graph To compare data that does not change continuously you might choose a bar graph. A bar graph uses bars to show the relationships between variables. The *x*-axis variable is divided into parts. The parts can be numbers such as years, or a category such as a type of animal. The *y*-axis is a number and increases continuously along the axis.

Example

A recycling center collects 4.0 kg of aluminum on Monday, 1.0 kg on Wednesday, and 2.0 kg on Friday. Create a bar graph of this data.

Step 1 Select the *x*-axis and *y*-axis variables. The measured numbers (the masses of aluminum) should be placed on the *y*-axis. The variable divided into parts (collection days) is placed on the *x*-axis.

Step 2 Create a graph grid like you would for a line graph. Include labels and units.

Step 3 For each measured number, draw a vertical bar above the *x*-axis value up to the *y*-axis value. For the first data point, draw a vertical bar above Monday up to 4.0 kg.

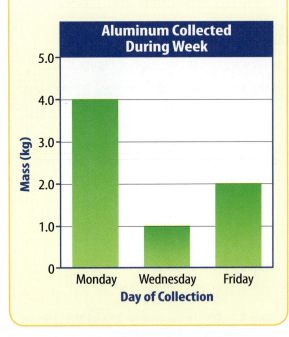

Practice Problem Draw a bar graph of the gases in air: 78% nitrogen, 21% oxygen, 1% other gases.

Circle Graph To display data as parts of a whole, you might use a circle graph. A circle graph is a circle divided into sections that represent the relative size of each piece of data. The entire circle represents 100%, half represents 50%, and so on.

Example

Air is made up of 78% nitrogen, 21% oxygen, and 1% other gases. Display the composition of air in a circle graph.

Step 1 Multiply each percent by 360° and divide by 100 to find the angle of each section in the circle.

$$78\% \times \frac{360°}{100} = 280.8°$$

$$21\% \times \frac{360°}{100} = 75.6°$$

$$1\% \times \frac{360°}{100} = 3.6°$$

Step 2 Use a compass to draw a circle and to mark the center of the circle. Draw a straight line from the center to the edge of the circle.

Step 3 Use a protractor and the angles you calculated to divide the circle into parts. Place the center of the protractor over the center of the circle and line the base of the protractor over the straight line.

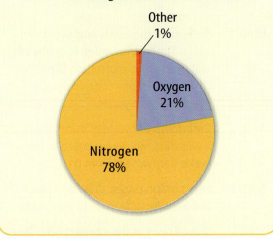

Practice Problem Draw a circle graph to represent the amount of aluminum collected during the week shown in the bar graph to the left.

SR-28 • Math Skill Handbook

FOLDABLES® Handbook

Student Study Guides & Instructions
By Dinah Zike

1. You will find suggestions for Study Guides, also known as Foldables or books, in each chapter lesson and as a final project. Look at the end of the chapter to determine the project format and glue the Foldables in place as you progress through the chapter lessons.

2. Creating the Foldables or books is simple and easy to do by using copy paper, art paper, and internet printouts. Photocopies of maps, diagrams, or your own illustrations may also be used for some of the Foldables. Notebook paper is the most common source of material for study guides and 83% of all Foldables are created from it. When folded to make books, notebook paper Foldables easily fit into 11" × 17" or 12" × 18" chapter projects with space left over. Foldables made using photocopy paper are slightly larger and they fit into Projects, but snugly. Use the least amount of glue, tape, and staples needed to assemble the Foldables.

3. Seven of the Foldables can be made using either small or large paper. When 11" × 17" or 12" × 18" paper is used, these become projects for housing smaller Foldables. Project format boxes are located within the instructions to remind you of this option.

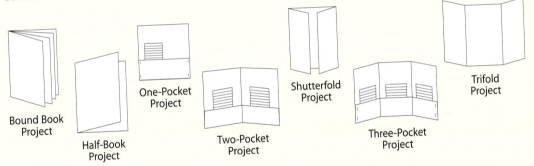

4. Use one-gallon self-locking plastic bags to store your projects. Place strips of two-inch clear tape along the left, long side of the bag and punch holes through the taped edge. Cut the bottom corners off the bag so it will not hold air. Store this Project Portfolio inside a three-hole binder. To store a large collection of project bags, use a giant laundry-soap box. Holes can be punched in some of the Foldable Projects so they can be stored in a three-hole binder without using a plastic bag. Punch holes in the pocket books before gluing or stapling the pocket.

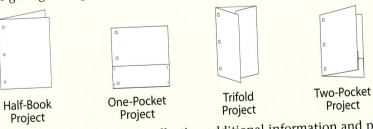

5. Maximize the use of the projects by collecting additional information and placing it on the back of the project and other unused spaces of the large Foldables.

Half-Book Foldable® By Dinah Zike

Step 1 Fold a sheet of notebook or copy paper in half.

Label the exterior tab and use the inside space to write information.

PROJECT FORMAT
Use 11" × 17" or 12" × 18" paper on the horizontal axis to make a large project book.

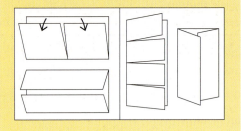

Variations
Paper can be folded horizontally, like a *hamburger* or vertically, like a *hot dog*.

C Half-books can be folded so that one side is ½ inch longer than the other side. A title or question can be written on the extended tab.

Worksheet Foldable or Folded Book® By Dinah Zike

Step 1 Make a half-book (see above) using work sheets, internet print-outs, diagrams, or maps.

Step 2 Fold it in half again.

Variations

A This folded sheet as a small book with two pages can be used for comparing and contrasting, cause and effect, or other skills.

B When the sheet of paper is open, the four sections can be used separately or used collectively to show sequences or steps.

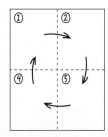

Two-Tab and Concept-Map Foldable® By Dinah Zike

Step 1 Fold a sheet of notebook or copy paper in half vertically or horizontally.

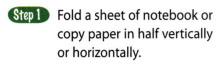

Step 2 Fold it in half again, as shown.

Step 3 Unfold once and cut along the fold line or valley of the top flap to make two flaps.

Variations

A Concept maps can be made by leaving a ½ inch tab at the top when folding the paper in half. Use arrows and labels to relate topics to the primary concept.

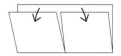

B Use two sheets of paper to make multiple page tab books. Glue or staple books together at the top fold.

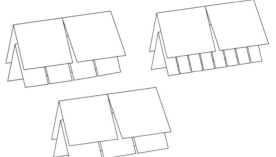

Three-Quarter Foldable® By Dinah Zike

Step 1 Make a two-tab book (see above) and cut the left tab off at the top of the fold line.

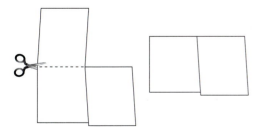

Variations

A Use this book to draw a diagram or a map on the exposed left tab. Write questions about the illustration on the top right tab and provide complete answers on the space under the tab.

B Compose a self-test using multiple choice answers for your questions. Include the correct answer with three wrong responses. The correct answers can be written on the back of the book or upside down on the bottom of the inside page.

Three-Tab Foldable® By Dinah Zike

Step 1 Fold a sheet of paper in half horizontally.

Step 2 Fold into thirds.

Step 3 Unfold and cut along the folds of the top flap to make three sections.

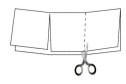

Variations

A Before cutting the three tabs draw a Venn diagram across the front of the book.

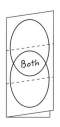

B Make a space to use for titles or concept maps by leaving a ½ inch tab at the top when folding the paper in half.

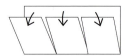

Four-Tab Foldable® By Dinah Zike

Step 1 Fold a sheet of paper in half horizontally.

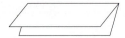

Step 2 Fold in half and then fold each half as shown below.

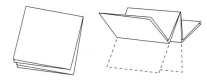

Step 3 Unfold and cut along the fold lines of the top flap to make four tabs.

Variations

A Make a space to use for titles or concept maps by leaving a ½ inch tab at the top when folding the paper in half.

B Use the book on the vertical axis, with or without an extended tab.

Folding Fifths for a Foldable® By Dinah Zike

Step 1 Fold a sheet of paper in half horizontally.

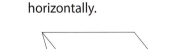

Step 2 Fold again so one-third of the paper is exposed and two-thirds are covered.

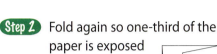

Step 3 Fold the two-thirds section in half.

Step 4 Fold the one-third section, a single thickness, backward to make a fold line.

Variations

A Unfold and cut along the fold lines to make five tabs.

B Make a five-tab book with a ½ inch tab at the top (see two-tab instructions).

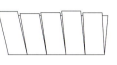

C Use 11" × 17" or 12" × 18" paper and fold into fifths for a five-column and/or row table or chart.

Folded Table or Chart, and Trifold Foldable® By Dinah Zike

Step 1 Fold a sheet of paper in the required number of vertical columns for the table or chart.

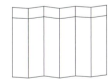

Step 2 Fold the horizontal rows needed for the table or chart.

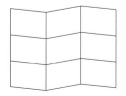

PROJECT FORMAT
Use 11" × 17" or 12" × 18" paper and fold it to make a large trifold project book or larger tables and charts.

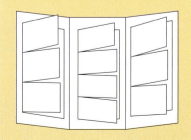

Variations

A Make a trifold by folding the paper into thirds vertically or horizontally.

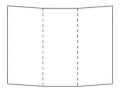

B Make a trifold book. Unfold it and draw a Venn diagram on the inside.

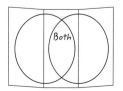

Two or Three-Pockets Foldable® By Dinah Zike

Step 1 Fold up the long side of a horizontal sheet of paper about 5 cm.

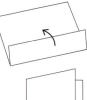

Step 2 Fold the paper in half.

Step 3 Open the paper and glue or staple the outer edges to make two compartments.

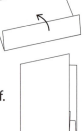

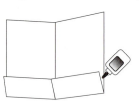

Variations

A Make a multi-page booklet by gluing several pocket books together.

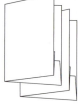

B Make a three-pocket book by using a trifold (see previous instructions).

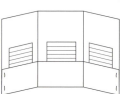

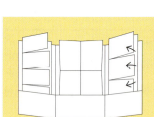

PROJECT FORMAT
Use 11" × 17" or 12" × 18" paper and fold it horizontally to make a large multi-pocket project.

Matchbook Foldable® By Dinah Zike

Step 1 Fold a sheet of paper almost in half and make the back edge about 1–2 cm longer than the front edge.

Step 2 Find the midpoint of the shorter flap.

Step 3 Open the paper and cut the short side along the midpoint making two tabs.

Step 4 Close the book and fold the tab over the short side.

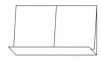

Variations

A Make a single-tab matchbook by skipping Steps 2 and 3.

B Make two smaller matchbooks by cutting the single-tab matchbook in half.

Shutterfold Foldable® By Dinah Zike

Step 1 Begin as if you were folding a vertical sheet of paper in half, but instead of creasing the paper, pinch it to show the midpoint.

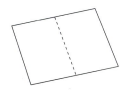

Step 2 Fold the top and bottom to the middle and crease the folds.

Variations

A Use the shutterfold on the horizontal axis.

B Create a center tab by leaving .5–2 cm between the flaps in Step 2.

PROJECT FORMAT
Use 11" × 17" or 12" × 18" paper and fold it to make a large shutterfold project.

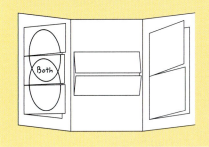

Four-Door Foldable® By Dinah Zike

Step 1 Make a shutterfold (see above).

Step 2 Fold the sheet of paper in half.

Step 3 Open the last fold and cut along the inside fold lines to make four tabs.

Variations

A Use the four-door book on the opposite axis.

B Create a center tab by leaving .5–2 cm between the flaps in Step 1.

Bound Book Foldable® By Dinah Zike

Step 1 Fold three sheets of paper in half. Place the papers in a stack, leaving about .5 cm between each top fold. Mark all three sheets about 3 cm from the outer edges.

Step 2 Using two of the sheets, cut from the outer edges to the marked spots on each side. On the other sheet, cut between the marked spots.

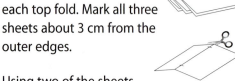

Step 3 Take the two sheets from Step 1 and slide them through the cut in the third sheet to make a 12-page book.

Step 4 Fold the bound pages in half to form a book.

Variation

A Use two sheets of paper to make an eight-page book, or increase the number of pages by using more than three sheets.

> **PROJECT FORMAT**
> Use two or more sheets of 11" × 17" or 12" × 18" paper and fold it to make a large bound book project.
>
>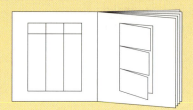

Accordian Foldable® By Dinah Zike

Step 1 Fold the selected paper in half vertically, like a *hamburger*.

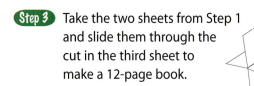

Step 2 Cut each sheet of folded paper in half along the fold lines.

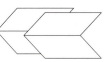

Step 3 Fold each half-sheet almost in half, leaving a 2 cm tab at the top.

Step 4 Fold the top tab over the short side, then fold it in the opposite direction.

Variations

A Glue the straight edge of one paper inside the tab of another sheet. Leave a tab at the end of the book to add more pages.

B Tape the straight edge of one paper to the tab of another sheet, or just tape the straight edges of nonfolded paper end to end to make an accordian.

C Use whole sheets of paper to make a large accordian.

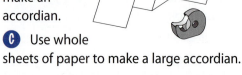

Layered Foldable® By Dinah Zike

Step 1 Stack two sheets of paper about 1–2 cm apart. Keep the right and left edges even.

Step 2 Fold up the bottom edges to form four tabs. Crease the fold to hold the tabs in place.

Step 3 Staple along the folded edge, or open and glue the papers together at the fold line.

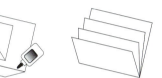

Variations

A Rotate the book so the fold is at the top or to the side.

B Extend the book by using more than two sheets of paper.

Envelope Foldable® By Dinah Zike

Step 1 Fold a sheet of paper into a *taco*. Cut off the tab at the top.

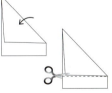

Step 2 Open the *taco* and fold it the opposite way making another *taco* and an X-fold pattern on the sheet of paper.

Step 3 Cut a map, illustration, or diagram to fit the inside of the envelope.

Step 4 Use the outside tabs for labels and inside tabs for writing information.

Variations

A Use 11″ × 17″ or 12″ × 18″ paper to make a large envelope.

B Cut off the points of the four tabs to make a window in the middle of the book.

Sentence Strip Foldable® By Dinah Zike

Step 1 Fold two sheets of paper in half vertically, like a *hamburger*.
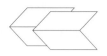

Step 2 Unfold and cut along fold lines making four half sheets.

Step 3 Fold each half sheet in half horizontally, like a *hot dog*.

Step 4 Stack folded horizontal sheets evenly and staple together on the left side.

Step 5 Open the top flap of the first sentence strip and make a cut about 2 cm from the stapled edge to the fold line. This forms a flap that can be raisied and lowered. Repeat this step for each sentence strip.
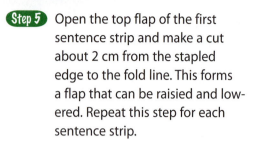

Variations

A Expand this book by using more than two sheets of paper.

B Use whole sheets of paper to make large books.

Pyramid Foldable® By Dinah Zike

Step 1 Fold a sheet of paper into a *taco*. Crease the fold line, but do not cut it off.

Step 2 Open the folded sheet and refold it like a *taco* in the opposite direction to create an X-fold pattern.

Step 3 Cut one fold line as shown, stopping at the center of the X-fold to make a flap.

Step 4 Outline the fold lines of the X-fold. Label the three front sections and use the inside spaces for notes. Use the tab for the title.

Step 5 Glue the tab into a project book or notebook. Use the space under the pyramid for other information.

Step 6 To display the pyramid, fold the flap under and secure with a paper clip, if needed.

Single-Pocket or One-Pocket Foldable® By Dinah Zike

Step 1 Using a large piece of paper on a vertical axis, fold the bottom edge of the paper upwards, about 5 cm.

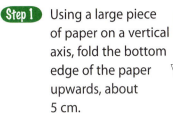

Step 2 Glue or staple the outer edges to make a large pocket.

Variations

A Make the one-pocket project using the paper on the horizontal axis.

B To store materials securely inside, fold the top of the paper almost to the center, leaving about 2–4 cm between the paper edges. Slip the Foldables through the opening and under the top and bottom pockets.

PROJECT FORMAT
Use 11" × 17" or 12" × 18" paper and fold it vertically or horizontally to make a large pocket project.

Multi-Tab Foldable® By Dinah Zike

Step 1 Fold a sheet of notebook paper in half like a *hot dog*.

Step 2 Open the paper and on one side cut every third line. This makes ten tabs on wide ruled notebook paper and twelve tabs on college ruled.

Step 3 Label the tabs on the front side and use the inside space for definitions or other information.

Variation

A Make a tab for a title by folding the paper so the holes remain uncovered. This allows the notebook Foldable to be stored in a three-hole binder.

Reference Handbook

PERIODIC TABLE OF THE ELEMENTS

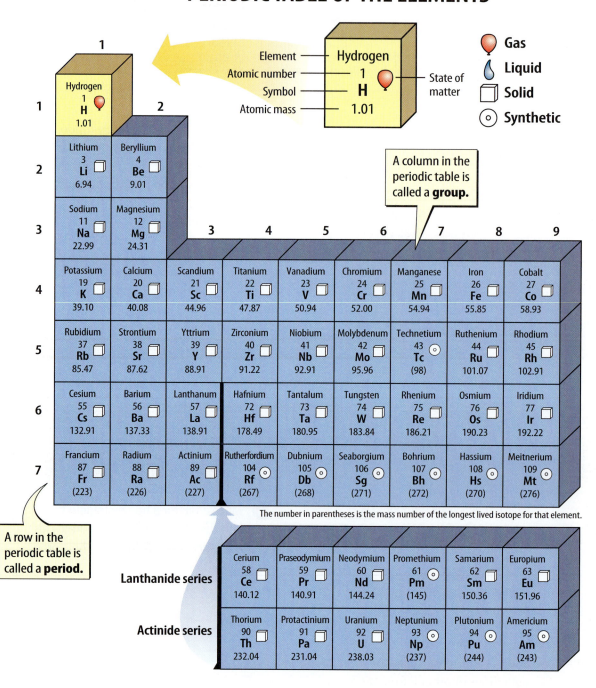

Legend:
- Metal
- Metalloid
- Nonmetal
- Recently discovered

13	14	15	16	17	18
					Helium 2 **He** 4.00
Boron 5 **B** 10.81	Carbon 6 **C** 12.01	Nitrogen 7 **N** 14.01	Oxygen 8 **O** 16.00	Fluorine 9 **F** 19.00	Neon 10 **Ne** 20.18
Aluminum 13 **Al** 26.98	Silicon 14 **Si** 28.09	Phosphorus 15 **P** 30.97	Sulfur 16 **S** 32.07	Chlorine 17 **Cl** 35.45	Argon 18 **Ar** 39.95

10	11	12	13	14	15	16	17	18
Nickel 28 **Ni** 58.69	Copper 29 **Cu** 63.55	Zinc 30 **Zn** 65.38	Gallium 31 **Ga** 69.72	Germanium 32 **Ge** 72.64	Arsenic 33 **As** 74.92	Selenium 34 **Se** 78.96	Bromine 35 **Br** 79.90	Krypton 36 **Kr** 83.80
Palladium 46 **Pd** 106.42	Silver 47 **Ag** 107.87	Cadmium 48 **Cd** 112.41	Indium 49 **In** 114.82	Tin 50 **Sn** 118.71	Antimony 51 **Sb** 121.76	Tellurium 52 **Te** 127.60	Iodine 53 **I** 126.90	Xenon 54 **Xe** 131.29
Platinum 78 **Pt** 195.08	Gold 79 **Au** 196.97	Mercury 80 **Hg** 200.59	Thallium 81 **Tl** 204.38	Lead 82 **Pb** 207.20	Bismuth 83 **Bi** 208.98	Polonium 84 **Po** (209)	Astatine 85 **At** (210)	Radon 86 **Rn** (222)
Darmstadtium 110 **Ds** (281)	Roentgenium 111 **Rg** (280)	Copernicium 112 **Cn** (285)	Ununtrium * 113 **Uut** (284)	Ununquadium * 114 **Uuq** (289)	Ununpentium * 115 **Uup** (288)	Ununhexium * 116 **Uuh** (293)		Ununoctium * 118 **Uuo** (294)

* The names and symbols for elements 113–116 and 118 are temporary. Final names will be selected when the elements' discoveries are verified.

Gadolinium 64 **Gd** 157.25	Terbium 65 **Tb** 158.93	Dysprosium 66 **Dy** 162.50	Holmium 67 **Ho** 164.93	Erbium 68 **Er** 167.26	Thulium 69 **Tm** 168.93	Ytterbium 70 **Yb** 173.05	Lutetium 71 **Lu** 174.97
Curium 96 **Cm** (247)	Berkelium 97 **Bk** (247)	Californium 98 **Cf** (251)	Einsteinium 99 **Es** (252)	Fermium 100 **Fm** (257)	Mendelevium 101 **Md** (258)	Nobelium 102 **No** (259)	Lawrencium 103 **Lr** (262)

Topographic Map Symbols

Topographic Map Symbols

Symbol	Description	Symbol	Description
	Primary highway, hard surface		Index contour
	Secondary highway, hard surface		Supplementary contour
	Light-duty road, hard or improved surface		Intermediate contour
	Unimproved road		Depression contours
	Railroad: single track		
	Railroad: multiple track		Boundaries: national
	Railroads in juxtaposition		State
			County, parish, municipal
	Buildings		Civil township, precinct, town, barrio
	Schools, church, and cemetery		Incorporated city, village, town, hamlet
	Buildings (barn, warehouse, etc.)		Reservation, national or state
	Wells other than water (labeled as to type)		Small park, cemetery, airport, etc.
	Tanks: oil, water, etc. (labeled only if water)		Land grant
	Located or landmark object; windmill		Township or range line, U.S. land survey
	Open pit, mine, or quarry; prospect		Township or range line, approximate location
	Marsh (swamp)		
	Wooded marsh		Perennial streams
	Woods or brushwood		Elevated aqueduct
	Vineyard		Water well and spring
	Land subject to controlled inundation		Small rapids
	Submerged marsh		Large rapids
	Mangrove		Intermittent lake
	Orchard		Intermittent stream
	Scrub		Aqueduct tunnel
	Urban area		Glacier
			Small falls
x7369	Spot elevation		Large falls
670	Water elevation		Dry lake bed

Rocks

Rocks		
Rock Type	Rock Name	Characteristics
Igneous (intrusive)	Granite	Large mineral grains of quartz, feldspar, hornblende, and mica. Usually light in color.
	Diorite	Large mineral grains of feldspar, hornblende, and mica. Less quartz than granite. Intermediate in color.
	Gabbro	Large mineral grains of feldspar, augite, and olivine. No quartz. Dark in color.
Igneous (extrusive)	Rhyolite	Small mineral grains of quartz, feldspar, hornblende, and mica, or no visible grains. Light in color.
	Andesite	Small mineral grains of feldspar, hornblende, and mica or no visible grains. Intermediate in color.
	Basalt	Small mineral grains of feldspar, augite, and possibly olivine or no visible grains. No quartz. Dark in color.
	Obsidian	Glassy texture. No visible grains. Volcanic glass. Fracture looks like broken glass.
	Pumice	Frothy texture. Floats in water. Usually light in color.
Sedimentary (detrital)	Conglomerate	Coarse grained. Gravel or pebble-size grains.
	Sandstone	Sand-sized grains 1/16 to 2 mm.
	Siltstone	Grains are smaller than sand but larger than clay.
	Shale	Smallest grains. Often dark in color. Usually platy.
Sedimentary (chemical or organic)	Limestone	Major mineral is calcite. Usually forms in oceans and lakes. Often contains fossils.
	Coal	Forms in swampy areas. Compacted layers of organic material, mainly plant remains.
Sedimentary (chemical)	Rock Salt	Commonly forms by the evaporation of seawater.
Metamorphic (foliated)	Gneiss	Banding due to alternate layers of different minerals, of different colors. Parent rock often is granite.
	Schist	Parallel arrangement of sheetlike minerals, mainly micas. Forms from different parent rocks.
	Phyllite	Shiny or silky appearance. May look wrinkled. Common parent rocks are shale and slate.
	Slate	Harder, denser, and shinier than shale. Common parent rock is shale.
Metamorphic (nonfoliated)	Marble	Calcite or dolomite. Common parent rock is limestone.
	Soapstone	Mainly of talc. Soft with greasy feel.
	Quartzite	Hard with interlocking quartz crystals. Common parent rock is sandstone.

Minerals

Minerals

Mineral (formula)	Color	Streak	Hardness Pattern	Breakage Properties	Uses and Other
Graphite (C)	black to gray	black to gray	1–1.5	basal cleavage (scales)	pencil lead, lubricants for locks, rods to control some small nuclear reactions, battery poles
Galena (PbS)	gray	gray to black	2.5	cubic cleavage perfect	source of lead, used for pipes, shields for X rays, fishing equipment sinkers
Hematite (Fe_2O_3)	black or reddish-brown	reddish-brown	5.5–6.5	irregular fracture	source of iron; converted to pig iron, made into steel
Magnetite (Fe_3O_4)	black	black	6	conchoidal fracture	source of iron, attracts a magnet
Pyrite (FeS_2)	light, brassy, yellow	greenish-black	6–6.5	uneven fracture	fool's gold
Talc ($Mg_3Si_4O_{10}(OH)_2$)	white, greenish	white	1	cleavage in one direction	used for talcum powder, sculptures, paper, and tabletops
Gypsum ($CaSO_4 \cdot 2H_2O$)	colorless, gray, white, brown	white	2	basal cleavage	used in plaster of paris and dry wall for building construction
Sphalerite (ZnS)	brown, reddish-brown, greenish	light to dark brown	3.5–4	cleavage in six directions	main ore of zinc; used in paints, dyes, and medicine
Muscovite ($KAl_3Si_3O_{10}(OH)_2$)	white, light gray, yellow, rose, green	colorless	2–2.5	basal cleavage	occurs in large, flexible plates; used as an insulator in electrical equipment, lubricant
Biotite ($K(Mg,Fe)_3(AlSi_3O_{10})(OH)_2$)	black to dark brown	colorless	2.5–3	basal cleavage	occurs in large, flexible plates
Halite (NaCl)	colorless, red, white, blue	colorless	2.5	cubic cleavage	salt; soluble in water; a preservative

Minerals

Minerals

Mineral (formula)	Color	Streak	Hardness	Breakage Pattern	Uses and Other Properties
Calcite ($CaCO_3$)	colorless, white, pale blue	colorless, white	3	cleavage in three directions	fizzes when HCl is added; used in cements and other building materials
Dolomite ($CaMg(CO_3)_2$)	colorless, white, pink, green, gray, black	white	3.5–4	cleavage in three directions	concrete and cement; used as an ornamental building stone
Fluorite (CaF_2)	colorless, white, blue, green, red, yellow, purple	colorless	4	cleavage in four directions	used in the manufacture of optical equipment; glows under ultraviolet light
Hornblende $(CaNa)_{2-3}(Mg,Al,Fe)_5-(Al,Si)_2Si_6O_{22}(OH)_2$	green to black	gray to white	5–6	cleavage in two directions	will transmit light on thin edges; 6-sided cross section
Feldspar ($KAlSi_3O_8$) ($NaAlSi_3O_8$), ($CaAl_2Si_2O_8$)	colorless, white to gray, green	colorless	6	two cleavage planes meet at 90° angle	used in the manufacture of ceramics
Augite $((Ca,Na)(Mg,Fe,Al)(Al,Si)_2O_6)$	black	colorless	6	cleavage in two directions	square or 8-sided cross section
Olivine $((Mg,Fe)_2SiO_4)$	olive, green	none	6.5–7	conchoidal fracture	gemstones, refractory sand
Quartz (SiO_2)	colorless, various colors	none	7	conchoidal fracture	used in glass manufacture, electronic equipment, radios, computers, watches, gemstones

Weather Map Symbols

Sample Station Model

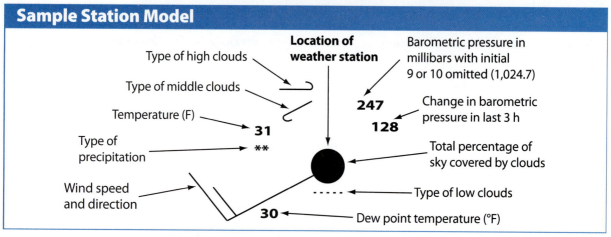

Sample Plotted Report at Each Station

Precipitation		Wind Speed and Direction		Sky Coverage		Some Types of High Clouds	
≡	Fog	○	0 calm	○	No cover	⌒	Scattered cirrus
★	Snow	/	1–2 knots	◐	1/10 or less	⌒⌒	Dense cirrus in patches
●	Rain	⌐	3–7 knots	◐	2/10 to 3/10	⌒⌒⌒	Veil of cirrus covering entire sky
⊺	Thunderstorm	⊢	8–12 knots	◐	4/10	⌒⌒	Cirrus not covering entire sky
'	Drizzle	⊧	13–17 knots	◐	–		
∇	Showers	⊨	18–22 knots	◐	6/10		
		⊫	23–27 knots	◐	7/10		
		⌐	48–52 knots	◐	Overcast with openings		
		1 knot = 1.852 km/h		●	Completely overcast		

Some Types of Middle Clouds		Some Types of Low Clouds		Fronts and Pressure Systems	
∠	Thin altostratus layer	⌒	Cumulus of fair weather	Ⓗ or High / Ⓛ or Low	Center of high- or low-pressure system
⫽	Thick altostratus layer	⌣	Stratocumulus	▲▲▲▲	Cold front
⌒∠	Thin altostratus in patches	-----	Fractocumulus of bad weather	⌒⌒⌒⌒	Warm front
⌒∠	Thin altostratus in bands	—	Stratus of fair weather	▲⌒▲⌒	Occluded front
				⌒▼⌒▼	Stationary front

Glossary/Glosario

Multilingual eGlossary

A science multilingual glossary is available on the science Web site. The glossary includes the following languages.

Arabic	Hmong	Tagalog
Bengali	Korean	Urdu
Chinese	Portuguese	Vietnamese
English	Russian	
Haitian Creole	Spanish	

Cómo usar el glosario en español:
1. Busca el término en inglés que desees encontrar.
2. El término en español, junto con la definición, se encuentran en la columna de la derecha.

Pronunciation Key

Use the following key to help you sound out words in the glossary.

a	back (BAK)		ew	food (FEWD)
ay	day (DAY)		yoo	pure (PYOOR)
ah	father (FAH thur)		yew	few (FYEW)
ow	flower (FLOW ur)		uh	comma (CAH muh)
ar	car (CAR)		u (+ con)	rub (RUB)
e	less (LES)		sh	shelf (SHELF)
ee	leaf (LEEF)		ch	nature (NAY chur)
ih	trip (TRIHP)		g	gift (GIHFT)
i (i + com + e)	idea (i DEE uh)		j	gem (JEM)
oh	go (GOH)		ing	sing (SING)
aw	soft (SAWFT)		zh	vision (VIH zhun)
or	orbit (OR buht)		k	cake (KAYK)
oy	coin (COYN)		s	seed, cent (SEED, SENT)
oo	foot (FOOT)		z	zone, raise (ZOHN, RAYZ)

English — A — Español

abyssal plains/aquifer — **planos abisales/acuífero**

abyssal plains: large, flat areas of the seafloor that extend across the deepest parts of ocean basins. (p. 566)

acid precipitation: precipitation that has a lower pH than that of normal rainwater (pH 5.6). (p. 670)

adhesion: the attraction among molecules that are not alike. (p. 539)

alpine glacier: a glacier that forms in the mountains. (p. 608)

aquifer: an area of permeable sediment or rock that holds significant amounts of water. (p. 627)

planos abisales: áreas extensas y planas del lecho marino que se extienden por las partes más profundas de las cuencas marinas. (pág. 566)

precipitación ácida: precipitación que tiene un pH más bajo que el del agua de la lluvia normal (pH 5.6). (pág. 670)

adhesión: atracción entre moléculas que son diferentes. (pág. 539)

glacial alpino: glacial que se forma en las montañas. (pág. 608)

acuífero: área de sedimento permeable o roca que conserva cantidades significativas de agua. (pág. 627)

bioindicator: an organism that is sensitive to environmental conditions and is one of the first to respond to changes. (p. 550)

biomass energy: energy produced by burning organic matter, such as wood, food scraps, and alcohol. (p. 655)

brackish water: a mix of fresh water and sea water. (p. 565)

bioindicador: organismo que es sensible a las condiciones medioambientales y es uno de los primeros en responder a los cambios. (pág. 550)

energía de biomasa: energía producida por la combustión de materia orgánica, como la madera, las sobras de comida y el alcohol. (pág. 655)

agua salobre: mezcla de agua dulce y agua de mar. (pág. 565)

cohesion: the attraction among molecules that are alike. (p. 539)

condensation: the process by which a gas changes to a liquid. (p. 531)

coral bleaching: the loss of color in corals that occurs when stressed corals expel the colorful algae that live in them. (p. 592)

Coriolis effect: the movement of wind and water to the right or left that is caused by Earth's rotation. (p. 582)

cohesión: atracción entre moléculas que son parecidas. (pág. 539)

condensación: proceso por el cual un organismo cambia a líquido. (pág. 531)

blanqueamiento de coral: pérdida de color en los corales que ocurre cuando los corales estresados expelen las algas de color que viven en ellos. (pág. 592)

efecto Coriolis: movimiento del viento y del agua a la derecha o a la izquierda causado por la rotación de la Tierra. (pág. 582)

deforestation: the removal of large areas of forests for human purposes. (p. 664)

deforestación: eliminación de grandes áreas de bosques con propósitos humanos. (pág. 664)

estuary: a coastal area where freshwater from rivers and streams mixes with salt water from seas or oceans. (p. 619)

evaporation: the process of a liquid changing to a gas at the surface of the liquid. (p. 531)

estuario: área costera donde el agua dulce de ríos y arroyos se mezcla con el agua salada de los mares u océanos. (pág. 619)

evaporación: proceso por el cual un líquido cambia a gas en la superficie de un líquido. (pág. 531)

freshwater: water that has less than 0.2 percent salt dissolved in it. (p. 607)

agua dulce: agua que tiene menos de 0,2 porciento de sal disuelta en ella. (pág. 607)

geothermal energy: thermal energy from Earth's interior. (p. 655)

energía geotérmica: energía térmica del interior de la Tierra. (pág. 655)

groundwater: water that is stored in cracks and pores beneath Earth's surface. (p. 625)

gyre: a large, circular system of ocean currents. (p. 582)

harmful algal bloom: explosive growth of algae that harms organisms. (p. 591)
hydroelectric power: electricity produced by flowing water. (p. 654)
hydrosphere: the system containing all of Earth's water. (p. 530)

ice core: a long column of ice taken from a glacier. (p. 612)
ice sheet: a glacier that spreads over land in all directions. (p. 609)

lake: a large body of water that forms in a basin surrounded by land. (p. 620)

marine: a term that refers to anything related to the oceans. (p. 590)

N

neap tide: the lowest tidal range that occurs when Earth, the Moon, and the Sun form a right angle. (p. 577)
nitrate: a nitrogen-based compound often used in fertilizers. (p. 549)
nonpoint-source pollution: pollution from several widespread sources that cannot be traced back to a single location. (p. 547)
nonrenewable resource: a resource that is used faster than it can be replaced by natural processes. (p. 643)
nuclear energy: energy stored in and released from the nucleus of an atom. (p. 647)

agua subterránea: agua almacenada en grietas o poros debajo de la superficie de la Tierra. (pág. 625)
giro: sistema circular extenso de corrientes marinas. (pág. 582)

floración de algas nocivas: crecimiento explosivo de algas dañinas para los organismos. (pág. 591)
energía hidroeléctrica: electricidad producida por agua que fluye. (pág. 654)
hidrosfera: sistema que contiene toda el agua de la Tierra. (pág. 530)

núcleo de hielo: columna larga de hielo tomado de un glacial. (pág. 612)
capa de hielo: glacial que se extiende sobre la tierra en todas las direcciones. (pág. 609)

lago: cuerpo extenso de agua que se forma en una cuenca rodeada de tierra. (pág. 620)

marino: término que se refiere a todo lo relacionado con los océanos. (pág. 590)

marea muerta: rango de marea más bajo que ocurre cuando la Tierra, la Luna y el Sol forman un ángulo recto. (pág. 577)
nitrato: compuesto con base en nitrógeno usado en los fertilizantes. (pág. 549)
contaminación de fuente no puntual: contaminación de varias fuentes apartadas que no se pueden rastrear hasta una sola ubicación. (pág. 547)
recurso no renovable: recurso que se usa más rápidamente de lo que se puede reemplazar mediante procesos naturales. (pág. 643)
energía nuclear: energía almacenada en y liberada por el núcleo de un átomo. (pág. 647)

ocean current/sea level   **corriente oceánica/nivel del mar**

O

ocean current: a large volume of water flowing in a certain direction. (p. 581)

ore: a deposit of minerals that is large enough to be mined for a profit. (p. 663)

corriente oceánica: gran cantidad de agua que fluye en cierta dirección. (pág. 581)

mena: depósito de minerales suficientemente grandes como para ser explotados con un beneficio. (pág. 663)

P

permeability: the measure of the ability of water to flow through rock and sediment. (p. 626)

photochemical smog: air pollution that forms from the interaction between chemicals in the air and sunlight. (p. 670)

point-source pollution: pollution from a single source that can be identified. (p. 547)

polarity: a condition in which opposite ends of a molecule have slightly opposite charges, but the overall charge of the molecule is neutral. (p. 538)

porosity: the measure of a rock's ability to hold water. (p. 626)

permeabilidad: medida de la capacidad del agua para fluir a través de la roca y el sedimento. (pág. 626)

smog fotoquímico: polución del aire que se forma de la interacción entre los químicos en el aire y la luz solar. (pág. 670)

contaminación de fuente puntual: contaminación de una sola fuente que se puede identificar. (pág. 547)

polaridad: condición en la cual los extremos opuestos de una molécula tienen cargas ligeramente opuestas, pero la carga completa de la molécula es neutra. (pág. 538)

porosidad: medida de la capacidad de una roca para almacenar agua. (pág. 626)

R

reclamation: a process in which mined land must be recovered with soil and replanted with vegetation. (p. 649)

remote sensing: the process of collecting information about an area without coming into contact with it. (p. 550)

renewable resource: a resource that can be replenished by natural processes at least as quickly as it is used. (p. 643)

runoff: water that flows over Earth's surface. (p. 617)

recuperación: proceso por el cual las tierras explotadas se deben recubrir con suelo y se deben replantar con vegetación. (pág. 649)

teledetección: proceso de recolectar información sobre un área sin entrar en contacto con ella. (pág. 550)

recurso renovable: recurso natural que se reabastece por procesos naturales al menos tan rápidamente como se usa. (pág. 643)

escorrentía: agua que fluye sobre la superficie de la Tierra. (pág. 617)

S

salinity: a measure of the mass of dissolved salts in a mass of water. (p. 565)

sea ice: ice that forms when sea water freezes. (p. 611)

sea level: the average level of the ocean's surface at any given time. (p. 576)

salinidad: medida de la masa de sales disueltas en una masa de agua. (pág. 565)

hielo marino: hielo que se forma cuando el agua del mar se congela. (pág. 611)

nivel del mar: promedio del nivel de la superficie del océano en algún momento dado. (pág. 576)

seawater: water from a sea or ocean that has an average salinity of 35 ppt. (p. 565)

solar energy: energy from the Sun. (p. 653)

specific heat: the amount of thermal energy (joules) needed to raise the temperature of 1 kg of material 1°C. (p. 529)

spring tide: the largest tidal range that occurs when Earth, the Moon, and the Sun form a straight line. (p. 577)

stream: a body of water that flows within a channel. (p. 618)

agua de mar: agua del mar o del océano que tiene una salinidad promedio de 35 ppt. (pág. 565)

energía solar: energía proveniente del Sol. (pág. 653)

calor específico: cantidad de energía térmica (julios) requerida para subir la temperatura de 1 kg de materia a 1°C. (pág. 529)

marea de primavera: rango de marea más alto que ocurre cuando la Tierra, la Luna y el Sol forman una línea recta. (pág. 577)

corriente: cuerpo de agua que fluye por un canal. (pág. 618)

tidal range: the difference in water level between a high tide and a low tide. (p. 576)

tide: the periodic rise and fall of the ocean's surface caused by gravitational force between Earth and the Moon, and Earth and the Sun. (p. 576)

transpiration: the process by which plants release water vapor through their leaves. (p. 533)

tsunami: a wave that forms when an ocean disturbance suddenly moves a large volume of water. (p. 575)

turbidity: a measure of the cloudiness of water from sediments, microscopic organisms, or pollutants. (p. 549)

rango de marea: diferencia en el nivel de agua entre una marea alta y una marea baja. (pág. 576)

marea: ascenso y descenso periódico de la superficie del océano causados por la fuerza gravitacional entre la Tierra y la Luna, y la Tierra y el Sol. (pág. 576)

transpiración: proceso por el cual las plantas liberan vapor de agua por medio de las hojas. (pág. 533)

tsunami: ola que se forma cuando una alteración en el océano mueve repentinamente una gran cantidad de agua. (pág. 575)

turbidez: medida de la turbiedad del agua debido a sedimentos, organismos microscópicos o contaminantes. (pág. 549)

upwelling: the vertical movement of water toward the ocean's surface. (p. 583)

surgencia: movimiento vertical del agua hacia la superficie del océano. (pág. 583)

water cycle: the series of natural processes by which water continually moves throughout the hydrosphere. (p. 532)

water quality: the chemical, biological, and physical status of a body of water. (p. 546)

ciclo del agua: serie de procesos naturales mediante la cual el agua se mueve continuamente en toda la hidrosfera. (pág. 532)

calidad del agua: estado químico, biológico y físico de un cuerpo de agua. (pág. 546)

water table: the upper limit of the underground region in which the cracks and pores within rocks and sediment are completely filled with water. (p. 626)

watershed: an area of land that drains runoff into a particular stream, lake, ocean or other body of water. (p. 619)

wetland: an area of land that is saturated with water for part or all of the year. (p. 628)

wind farm: a group of wind turbines that produce electricity. (p. 654)

nivel freático: límite superior de la región subterránea en la cual las grietas y los poros dentro de las rocas y el sedimento están completamente llenos de agua. (pág. 626)

cuenca hidrográfica: área de tierra que drena escorrentía hacia un arroyo, lago, océano u otro cuerpo de agua en particular. (pág. 619)

lago: área de tierra saturada con agua durante parte del año o todo el año. (pág. 628)

parque eólico: grupo de turbinas de viento que produce electricidad. (pág. 654)

Index

Abyssal plain(s)   *Italic numbers* = illustration/photo   **Bold numbers** = vocabulary term   Food web(s)
lab = indicates entry is used in a lab on this page

A

Abyssal plain(s), *566,* **566**
Academic Vocabulary, 540, 567, 621, 649. *See also* **Vocabulary**
Acid precipitation, 670
Acidity
  of water, 549
Adhesion, *539,* **539**
Agriculture
  land use for, 662, *662,* 665
Air pollution. *See also* **Pollution**
  management of, 672, *672*
  sources of, 670, *671, 672*
Air
  as resource, 660, 670, 672
Algae
  calcium use by, 594
  decay of, 548, 549
  decomposition of, 591
  effect of bacteria on, 621
  overgrowth of, 550
Algal bloom, 549, *549, 591,* **591**
Alpine glacier(s)
  explanation of, *608,* **608**
  melting of, 613, *613*
Antarctica
  ice sheets in, *608,* 609, *609,* 610
Aquifer(s), *627,* **627**
Arctic Ocean
  explanation of, 563
  sea ice in, 611, 613, *613,* 615
Asteroid(s)
  as source of ocean formation, 564
Atlantic Ocean, 563
Atomic Energy Act, 649

B

Basin(s), 620
Bauxite, 663
Bay of Fundy, 579
Bias
  identification of, 651
Big Idea, 524, 554, 560, 598, 604, 634, 640, 676
  Review, 557, 601, 637, 679
Bioindicators, 550
Biomass energy
  advantages and disadvantages of, 656
  explanation of, **655**
Body temperature, 528. *See also* **Temperature**
  role of water in, 528
Bog(s), *628*
Brackish water, 565
Breaker(s), 575
Brook(s), 618

C

Calcium
  in seawater, 594, *594,* 594 *lab*
California Current, 584, *584*
Carbon dioxide
  in atmosphere, 528, 612, *612*
  deforestation and, 664
  fossil fuel producing, 612
  in seawater, 593, *593*
Carbonic acid, 593
Careers in Science, 535, 571
Cargo spills, 587
Chapter Review, 556–557, 600–601, 636–637, 678–679, 672, *672*
Climate change
  effect of melting snow or sea ice on, 611, *612*
  human impact on, 612
  oceans and, 592, 592–594, *593, 594*
  sea level change and, 610
Cloud(s)
  explanation of, 530
  formation of, 564
Coal
  disadvantages of, 646, *646*
  explanation of, 644
  formation of, *644*
Cohesion, *539,* **539**
Comet(s)
  as source of ocean formation, 564
Common Use. *See* **Science Use v. Common Use**
Community gardens, *665*
Compost, 665
Condensation
  explanation of, **531,** 533
  as source of ocean formation, 564
Continental glaciers. *See* **Ice sheets**
Continental margin, 566, *566*
Continental shelf, 566, *566*
Continental slope, 566, *566*
Coral bleaching, 548, 592, *592*
Coral
  calcium absorption by, 594
  effect or temperature change on, 592, *592*
  information from fossilized, 535
Coriolis effect, 582
Creeks, 618
Critical thinking, 534, 543, 551, 557, 570, 578, 586, 595, 601, 614, 622, 631, 637, 650, 658, 666, 673, 679
Current(s). *See* **Ocean current(s)**

D

Decomposition
  of algae, 591
Deep zone, 568
Deep-sea vent(s), 571
Deforestation, 664
Density
  explanation of, 540
  seawater, 569
  water, *540,* 540–542, *541*
Density current(s), 583
Distinct, 621
DSV Alvin, 567

E

Earth
  distribution of water on, 530, *530*
  percent of water on, 607, 607 *lab*
  surface of, 625 *lab*
East Antarctic Ice Sheet, 609, *609,* 610, *610*
Ecosystem(s)
  in Arctic regions, 615
Energy. *See also* **Nonrenewable energy resources; Renewable energy resources**
  analyzing use of, 659
  efficient use of, 674–675 *lab*
  sources of, 643, *643,* 649, 657
  vampire, 649
Energy conservation
  methods for, 649
Energy Policy Act, 649
Erosion
  causes of, 590
  wetlands as protection against, 629
Estuaries, *619,* **619**
Evaporation
  explanation of, *531,* **531**
  of water from plants, 533

F

Fertilizer(s)
  runoff from, 549
Filtration systems
  wetlands as, 629
Floods
  wetlands as protection against, 629
Florida Current, 582
Foldables, 532, 539, 547, 555, 566, 575, 584, 593, 599, 610, 620, 627, 635, 649, 655, 664, 670, 677
Food web(s)
  seawater acidity and, 563

I-2 • Index

Forest(s), 662, *662*
Forest fires
  pollution from, 670
Fossil fuel(s)
  advantages of, 646
  carbon dioxide released from burning, 612, *612*
  coal as, 644, *644*
  disadvantages of, 646, *647*, 67
  explanation of, 644
  oil and natural gas as, 645, *645*
Freshwater
  density of, *541*, 542
  distribution of, 530
  distribution on Earth, *671*
  on Earth, 607, *607*
  explanation of, **607**
  in glaciers, 610
  in ice sheets, 609
  mixing with seawater, 619

Gas
  water vapor as, 531
Geologist(s)
  function of, 535
Geothermal energy
  advantages and disadvantages of, *656*
  explanation of, *655*, **655**
Glacial meltwater, 608
Glacier(s)
  alpine, 608, *608*
  explanation of, **608**
  freshwater in, 610
  human impact on, 612–613
  ice sheets as, *608*, 609, *609*
  melting of, 610, *610*, 613, *613*
Gravitational force
  between Earth and Moon, 576
Great Ocean Conveyor Belt, *585*, **585**
Green building(s), 667
Green Science, 667
Greenhouse effect, *528*, **528**
Greenhouse gas(es)
  in atmosphere, 528
Greenland
  ice sheets in, 535, *608*, 609
Greensburg, Kansas, 667
Groundwater
  explanation of, 530, **625**
  flow of, 626–627, *627*
  human impact on, 627
  importance of, 625
  water table and, 626, *626*
Gulf Stream
  surface currents in, 584, *584*
Gyre(s)
  explanation of, *582*, **582**
  ocean pollution and, 590, *590*

Habitat(s)
  effect of mining on, 646
  in wetlands, 629, *629*

Harmful algal bloom(s), 591
Hazardous waste
  from nuclear power plants, 648
Headwater(s), 619
Hematite, 663
Human being(s)
  percent of water in, 527
Human body
  water and oxygen in, 669
Humphris, Susan, 571
Hurricane Floyd
  runoff following, 550, *550*
Hurricane Katrina
  flooding following, 630
Hydroelectric power, *654*, **654**
Hydrosphere, 530, 531

Ice
  density of, 541, *541*, 542, 544
  explanation of, 530
Ice core
  explanation of, 612
  gas bubbles in, *612*
Ice sheet(s)
  explanation of, **609**
  location of, *608*, 609, *609*, 610
  melting of, 535
Indian Ocean, 563
Interpret Graphics, 534, 543, 551, 570, 578, 586, 595, 614, 622, 631, 650, 658, 666, 673

Jellyfish, 527, *527*

Key Concepts, 526, 536, 545, 562, 572, 580, 588, 591, 606, 616, 624, 642
  Check, 529, 530, 533, 537, 539, 542, 547, 550, 565, 567, 569, 574, 575, 576, 581, 583, 584, 585, 590, 592, 594, 610–612, 618, 619, 621, 625, 627, 629, 643, 647, 649, 655, 656, 661, 664, 665, 672
  Summary, 554, 598, 634, 676
  Understand, 534, 543, 556, 570, 578, 586, 595, 600, 614, 622, 631, 636, 650, 658, 666, 673, 678

Lab, 552–553, 596–597, 632–633, 674–675. *See also* **Launch Lab; MiniLab; Skill Practice**
Lake(s)
  explanation of, **620**
  formation of, 620
  human impact on, 621, *621*
  pollution in, 620 *lab*
  properties and structure of, 621
Land resource(s)
  deforestation of, 664, *664*

  explanation of, 661, *661*
  forests as, 662–663, *663*
  management of, 662 *lab,* 665, *665*
  pollution of, 664
  in United states, *661*, 662
Launch Lab, 527, 537, 546, 563, 573, 581, 589, 607, 617, 625, 643, 653, 661, 669
Lava
  density of, 540
Lesson Review, 534, 543, 551, 570, 578, 586, 595, 614, 622, 631, 658, 666, 673
  explanation of, 650
Light
  in oceans, 568, 569
Liquid state
  water in, 531
Living space, 661
Louisiana
  wetlands in, 630, *630*

Macroinvertebrate(s)
  in streams, 617 *lab*
Magma
  thermal energy from, 655
Manipulator, 567
Mariana Trench, 567
Marine, 590
Marine geochemist(s), 571
Marsh(es), *628*
Math Skills, 529, 534, 557, 576, 601, 609, 614, 637, 672, 673, 679
Methane hydrate(s)
  from ocean floor, *567*
Methane
  in atmosphere, 528
Middle zone
  of ocean, 568, *568*
Mid-ocean ridge(s)
  explanation of, 566, *566*
Mineral(s)
  on ocean floor, *567*
Mineral resources
  common items from, 663, *663*
  mining for, 665
MiniLab, 531, 541, 548, 565, 577, 585, 594, 611, 620, 630, 647, 657, 662, 671. *See also* **Lab**
Moon
  gravitational force between Earth and, 576
  tides and position of, 577, *577*
Muhs, Daniel, 535

Natural gas
  explanation of, 645, *645*
  from ocean floor, *567*
Neap tide(s), 577, *577*
Nitrate(s)
  explanation of, **549**
  sources of excess, 591, *591*

**Nonpoint-source pollution, 547
Nonrenewable resource(s)**
  explanation of, *643*, **643**
  fossil fuels as, *644*, 644–647, *645*, *646*, *647*
  management of, 649
  nuclear energy as, 647–648, 647 *lab*, *648*
North Pacific Gyre, 590, *590*
North Pole
  sea ice in, 613, *613*
Nuclear energy
  647–648, 647 *lab*, *648*
Nuclear fission, 647 *lab*, *648*
Nuclear power plant(s)
  hazardous waste from, 648

O

Ocean(s). *See also* **Seafloor; Seawater**
  climate change and, *592*, 592–594, *593*, *594*
  composition of, 565, *565*
  explanation of, 563, *563*
  formation of, 564
  need for healthy, 594
  pollution of, *589*, 589–591, *590*, *591*
  seafloor of, *566*, 566–567, *567*
  trenches in, 567
  zones in, *568*, 568–569
Ocean current(s)
  density, 583
  explanation of, **581**
  Great Ocean Conveyor Belt and, 585, *585*
  research on, 587
  surface, *581*, 581–584, *582*, *583*
  weather and climate and, *584*, 584–585, *585*
Ocean wave(s)
  measurement of, *573*, 573
  parts of, *573*, 573
  surface, *574*, 574–575, *575*
  tsunamis as, 575
Oceanographers, 587
Office of Efficiency and Renewable Energy (Department of Energy), 657
Oil
  explanation of, 645, *645*
  from ocean floor, 567
Ore(s), 663
Oxygen
  dissolved, 548, 548 *lab*
  in human body, 669, *669*
  in water from riffles, 618

P

Pacific Ocean, 563
Pamlico Sound, 550, *550*
Permeability, 626
Petroleum. *See* **Oil**
Phosphate(s)
  sources of excess, 591

Photochemical smog, 670
Photosynthesis, 528, **568**, 664
Plankton, 645
Point-source pollution, 547
Polar, 538
Polar bears
  survival of, 615
Polarity
  explanation of, **538**
  of water molecules, 538, *538*
Pollution. *See also* **Air pollution; Water pollution**
  chemical waste as source of, 590
  efforts to avoid freshwater, 632–633 *lab*
  from fossil fuels, 647, *647*
  in groundwater, 627
  nonpoint-source, **547,** 589
  ocean, *589*, 589–591, *590*, *591*
  point-source, **547,** 589
  from runoff, 664
  in streams and lakes, 621, *621*
Pool(s), 618
Porosity, 626
Positive feedback loop, 613
Precipitation
  as source of ocean formation, 564
  effects of, 533
  explanation of, 533
Preserve, 665

R

Rain. *See also* **Precipitation**
  explanation of, 533
Reading Check, 528, 531, 532, 540, 541, 546, 548, 549, 564, 568, 573, 577, 582, 591, 593, 607, 609, 617, 626, 644, 645, 646, 648, 654, 662, 669, 670, 671
Reclamation, 649
Recycling
  benefits of, 665
Regulation
  of nuclear emissions, **649**
Remote sensing
  explanation of, 550
Remotely operated vehicles (ROVs), 567
Renewable energy resource(s)
  advantages and disadvantages of, 656, *656*
  biomass energy as, 655
  explanation of, *643*, **643**
  geothermal energy as, 655, *655*
  in home, 653 *lab*
  management of, 657, 667
  at school, 657 *lab*
  solar energy as, 653, *653*
  water energy as, 654, *654*
  wind energy as, 654, *654*
Reservoir(s), 533
Review Vocabulary, 540, 568, 608, 664. *See also* **Vocabulary**
Riffle(s), 618, *618*
Rill(s), 618

Rock(s)
  permeability of, 626
  porosity of, 626
Runoff
  effects of, 533
  explanation of, **617, 664**
  from fertilizers, 549
  pollutants in, 621, *621*
  of sediment, 550, *550*

S

Salinity
  explanation of, **565**
  in oceans, 569
  water density and, 565 *lab*
Salt water
  on Earth, 607
  sea ice as frozen, 611
Science & Society, 615
Science Methods, 553, 597, 675
Science Use v. Common Use, 538, 591, 620. *See also* **Vocabulary**
Sea ice
  explanation of, **611**
  location of, *613*
  melting of, 613
  polar bears and, 615
Sea level
  changes in, 610
  explanation of, **576**
  measurement of, 573 *lab*, 576
  melting of ice sheets and, 535
  sea ice and, 611
  temperature increase and, 592
Seafloor. *See also* **Oceans**
  exploration of, 567
  resources from, 567, *567*
  topography of, *566*, 566–567
  waves in contact with, 575, *575*
Seawater. *See also* **Oceans**
  acidity of, *593*, 593–594, 594 *lab*
  carbon dioxide in, 593, *593*
  composition of, 565, *565*
  density of, 569
  explanation of, **565**
  O_2 in, 592
  salinity of, 565, 569
  sunlight in, 568
  temperature of, 569, 592
Sediment
  as source of ocean pollution, 590, *590*
  permeability of, 626
  in water, 546 *lab*
Skill Practice, 544, 579, 587, 623, 651, 659. *See also* **Lab**
Smog, 670
Snow cover, 611
Snow
  relationship between rising temperature and melting, 613
Sodium chloride
  dissolved in water, 538
Solar cell(s), 653

Solar energy
  advantages and disadvantages of, 656
  explanation of, 643, 653, **653**
Solar power plants, 653
Solid state
  water in, 531
Solid waste
  as source of ocean pollution, 590, *590*
Solvent(s)
  water as, 539
Southern Ocean
  explanation of, 563
  wind-driven waves in, 574
Specific heat, 529
Sphagnum, 628
Spring(s), 627
Spring tide(s), 577, *577*
Standardized Test Practice, 558–559, 602–603, 638–639, 680–681
Stoneflies, 550
Stream(s)
  explanation of, *618*, **618**
  human impact on, 621, *621*
  that begin at headwaters, 619
  water flow into, 623
  water quality in, 617 *lab*
Strip-mining, 646, *646*
Study Guide, 554–555, 598–599, 634–635, 676–677
Sun
  tides and position of, 577, *577*
Sunlight
  in ocean, 568
Surface current(s)
  Coriolis effect and, 582, *582*
  effect on United States of, 584, *584*
  explanation of, 581, *581*
  gyres and, 582, *582*
  topography and, 582
  upwelling and, 583, *583*, 585
Surface wave(s). *See also* **Ocean waves**
  explanation of, 574
  motion of, 574, *574*
  that reach shore, 575, *575*
Surface zone
  of ocean, 568, *568*
Swamp(s), 628

T

Tectonic plate(s)
  ocean formation by movement in, 564
Temperature. *See also* **Body temperature**
  changes in water, 548
  effect of ground color on, 611 *lab*
  melting snow and rising, 613
  in oceans, 569
  oceans and increase of surface, 592, *592*
  stability of Earth's, 529
  states of water and, 531, 531 *lab*

  relationship between carbon dioxide and, 612, *612*
  of water, 621
  water density and, 541, *541*, 542, *542*, 552–553 *lab*
Thermal energy
  effects on water, 531
  explanation of, 529
  ocean currents and, 584, 585, 594
  from Sun, 528
Thermocline
  effect on lakes of, 620 *lab*
Tidal power, 654
Tidal range
  in Bay of Fundy, 579
  calculation of, 577 *lab*
  explanation of, **576**, 577
Tide(s)
  in Bay of Fundy, 579
  explanation of, **576**
  moon and, 576
  neap, 577
  spring, 577, *577*
  topography and, 576, *576*
Topography
  surface currents and, 582
  tides and, 576
Transpiration, 533
Tsunamis, 575
Turbidity
  explanation of, **549**
  measurement of, *549*

U

Upwelling
  explanation of, 583, *583*, 585
  whale sightings based on, 596–597 *lab*
Uranium, 648

V

Vampire energy, 649
Visual Check, 530, 533, 540, 542, 547, 550, 565, 566, 569, 574, 584, 613, 645, 649, 656, 662, 663, 670
Vocabulary, 525, 526, 536, 545, 554, 561, 562, 572, 580, 588, 598, 605, 606, 616, 624, 634, 642, 652, 660, 668. *See also* **Academic Vocabulary; Review Vocabulary; Science Use v. Common Use; Word Origin**
  Use, 534, 543, 551, 555, 570, 578, 586, 595, 599, 614, 622, 631, 635, 650, 658, 666, 673, 677
Volcanic eruptions
  explanation of, 564, *564*
  pollution from, 670, *670*

W

Water cycle
  explanation of, **532**
  paths in, *532*, 532–533

  role of runoff in, 617
  storage areas and, 533
Water density
  explanation of, 540
  features of, *540*, 542, 544
  temperature and, 541, *541*, 542, *542*, 552–553 *lab*
Water energy
  advantages and disadvantages of, 656
  explanation of, 654, *654*
Water molecules
  adhesion among, 539, *539*
  cohesion among, 539, *539*
  forces between, 537, *537*
  polarity of, 538, *538*
  as solvent, 538
Water pollution. *See also* **Pollution**
  nonpoint-source, 547, *547*
  point-source, 547, *547*
  sources of, 671
Water quality
  acidity and, 549
  bioindicators for, 550
  dissolved oxygen and, 548
  explanation of, **546**
  human effects on, 546
  nitrates and, 549, *549*
  turbidity and, 549, *549*
  water temperature and, 548
Water table, *626*, **626**
Water vapor
  in atmosphere, 528
  biological functions of, 527–528, *529*
  cloudy, 546 *lab*
  condensation of, 533, 564
  conservation of, 667
  distribution of, 530, *530*, 671
  explanation of, 531
  in human body, 669, *669*
  importance of, 527, 529
  properties of, 537, 537 *lab*
  as resource, 660, 671, 672
  role of wind in movement of, 581 *lab*
  specific heat of, 529
  states of, 531–533
  temperature of, 529
  temperature change in, 621
Watershed(s), *619*, **619**
Waves. *See* **Ocean waves**
Weather
  ocean currents and, 584, *584*, 585, *585*
Wells, 627
West Antarctic Ice Sheet, 609, *609*, 610, *610*
Wetland(s)
  explanation of, **628**
  human impact on, 630, *630*
  importance of, 629, *629*
  types of, 628, *628*
Whales
  upwelling and, 596–597 *lab*
What do you think?, 525, 534, 543, 551, 561, 570, 578, 586, 595, 605,

Wind energy
614, 622, 631, 641, 650, 658, 666, 673
Wind energy
advantages and disadvantages of, *656*
explanation of, 654
Wind farm(s), *654,* **654**
Wind
movement of water by, 581, 581 *lab*
surface currents and, 581
surface waves and, 574
Word Origin, 530, 539, 549, 566, 575, 582, 590, 608, 619, 643, 655, 663. *See also* **Vocabulary**
Writing In Science, 557, 601, 637, 679

Credits

Photo Credits

Front Cover blickwinkle/Alamy; **Spine-Back Cover** Walter Geiersperger/CORBIS; **Connect Ed** (t)Richard Hutchings, (c)Getty Images, (b)Jupiter Images/ThinkStock/Alamy; **i** Thinkstock/Getty Images; **iv** Ransom Studios **viii–ix** The McGraw-Hill Companies; **ix** (b)Fancy Photography/Veer **524–525** Gallo Images/CORBIS; **526** Chris Newbert/Minden Pictures; **527** (t)Hutchings Photography/Digital Light Source; (b)Digital Vision/Getty Images; **529** (t)altrendo nature/Getty Images; (c)Stocktrek Images/Getty Images; (b)Neil Emmerson/Getty Images; **531** (t)Hutchings Photography/Digital Light Source; **531** (b)The McGraw-Hill Companies, Inc./Stephen Frisch, photographer; **534** (t)Chris Newbert/Minden Pictures; (b)Neil Emmerson/Getty Images; **535** (t)American Museum of Natural History; (c)American Museum of Natural History; (b)American Museum of Natural History; **536** Momatiuk - Eastcott/CORBIS; **537** (t)Hutchings Photography/Digital Light Source; (b)PhotoLink/Getty Images; **538** Hutchings Photography/Digital Light Source; **539** (l)Visuals Unlimited/CORBIS; (r)Lester V. Bergman/CORBIS; **540** Ralph A. Clevenger/CORBIS; **541** Hutchings Photography/Digital Light Source; **543** Visuals Unlimited/CORBIS; **544** (all)Hutchings Photography/Digital Light Source; **545** Rich Reid/Getty Images; **546** (t)Hutchings Photography/Digital Light Source; (b)James Leynse/CORBIS; **547** (t)Photofusion Picture Library/Alamy; (b)Mark Tomalty/Masterfile; **548** DK Limited/CORBIS; **549** (t)Nick Hawkes; Ecoscene/CORBIS; (bl,br)Hutchings Photography/Digital Light Source; **550** (t,b)NASA images courtesy the MODIS Rapid Response Team.; **551** (t)Mark Tomalty/Masterfile; (c)Nick Hawkes; Ecoscene/CORBIS; (b)Hutchings Photography/Digital Light Source; **552** (6)Macmillan/McGraw-Hill; (others)Hutchings Photography/Digital Light Source; **553** Hutchings Photography/Digital Light Source; **554** (t)Digital Vision/Getty Images; (b)Mark Tomalty/Masterfile; **557** Gallo Images/CORBIS; **560–561** Frank Krahmer/Getty Images; **562** DAVID DOUBILET/National Geographic Stock; **563** (t)Hutchings Photography/Digital Light Source; (b)Image by Reto Stockli, NASA/Goddard Space Flight Center. Enhancements by Robert Simmon.; **564** (t)Steve Kaufman/CORBIS; (b)Pixtal/SuperStock; **565** Hutchings Photography/Digital Light Source; **567** (t)Arnulf Husmo/Getty Images; (c)KAZUHIRO NOGI/AFP/Getty Images, (b)Peter Ryan/Photo Researchers, Inc.; **568** (tl)Guillen Photography/UW/USA/Gulf of Mexico/Alamy; (tr)David Fleetham/Getty Images, (b)Stuart Westmorland/The Image Bank/Getty Images; **570** (t)Steve Kaufman/CORBIS; (b)Stuart Westmorland/The Image Bank/Getty Images; **571** (c)Susan Humphris; (r)American Museum of Natural History; **571** (bkgd)Ralph White/CORBIS; **572** Brian Bielmann; **573** Hutchings Photography/Digital Light Source; **576** Bill Brooks/Alamy; **578** Bill Brooks/Alamy; **580** Jacques Descloitres, MODIS Rapid Response Team, NASA/GSFC; **581** (t)Hutchings Photography/Digital Light Source, (b)Peter Sterling/Getty Images; **585** Hutchings Photography/Digital Light Source; **587** (t)Hutchings Photography/Digital Light Source; **587** (c,b)The McGraw-Hill Companies **588** Chris Cheadle/age fotostock; **589** T. O'Keefe/PhotoLink/Getty Images; **590** (t)Matt Cramer; (b)NASA; **591** (tl)LOOK Die Bildagentur der Fotografen GmbH/Alamy; (tr)Visuals Unlimited/CORBIS; (b)JLP/CORBIS; **592** (tl)Darryl Leniuk/Getty Images; (b)Timothy G. Laman/National Geographic/Getty Images; **593** (t)Steve Allen/Brand X/CORBIS; (b)National Geographic Society; **594** STEVE GSCHMEISSNER/SCIENCE PHOTO LIBRARY; **595** (tl)JLP/CORBIS; (cl)Timothy G. Laman/National Geographic/Getty Images; (bl)STEVE GSCHMEISSNER/SCIENCE PHOTO LIBRARY; (br)Steve Allen/Brand X/CORBIS; **596** (t to b)Macmillan/McGraw-Hill; (2)Heimir Harðar/Peter Arnold Inc.; (3)Biosphoto/Swann Christopher/Peter Arnold Inc.; (4)FLIP NICKLIN/MINDEN PICTURES/National Geographic Stock; (5)Digital Vision/PunchStock; **597** PATRICIO ROBLES GIL/MINDEN PICTURES/National Geographic Stock; **598** (t)DAVID DOUBILET/National Geographic Stock; (cl)Jacques Descloitres, MODIS Rapid Response Team, NASA/GSFC; (cr)Brian Bielmann; (b)Timothy G. Laman/National Geographic/Getty Images; **600** Bill Brooks/Alamy; **601** Frank Krahmer/Getty Images; **604–605** Jon Spaull/Getty Images; **606** Paul Souders/CORBIS; **607** Hutchings Photography/Digital Light Source; **608** (t)Ron Watts/Getty Images; (b)Yann Arthus-Bertrand/CORBIS; **611** (l)Janet Foster/Masterfile; (r)Hutchings Photography/Digital Light Source; **612** (t)Nick Cobbing/Alamy; (c)VIN MORGAN/AFP/Getty Images; (b)National Geographic/Getty Images; **613** (t)William O. Field, National Snow and Ice Data Center; (b)Bruce F. Molnia, U.S. Geological Survey; **615** (t.c)NASA/Goddard Space Flight Center Scientific Visualization Studio-Rob Gerston (GSFC), (b)Creatas/PunchStock; **616** ThinkStock/SuperStock; **617** Garry McMichael/Photo Researchers, Inc.; **618** Andreas Strauss/Getty Images; **620** Hutchings Photography/Digital Light Source; **622** Garry McMichael/Photo Researchers, Inc.; **623** (2)Macmillan/McGraw-Hill; (others)Hutchings Photography/Digital Light Source; **624** Jason Edwards/Getty Images; **625** Hutchings Photography/Digital Light Source; **628** (t)Brand X Pictures/PunchStock; (c)Panoramic Images/Getty Images; (b)Custom Life Science Images/Alamy; **629** Frank Krahmer/Masterfile; **630** (t)Hutchings Photography/Digital Light Source; (b)Kevin Fleming/CORBIS; **631** (t)Jason Edwards/Getty Images; (c)Brand X Pictures/PunchStock; (b)Frank Krahmer/Masterfile; **632** (3)Macmillan/McGraw-Hill; (others)Hutchings Photography/Digital Light Source; **633** Paul S. Howell/Liaison/Getty Images; **634** (t)Janet Foster/Masterfile; (c)Andreas Strauss/Getty Images; (b)Brand X Pictures/PunchStock; **636** Panoramic Images/Getty Images; **637** (l)Ron Watts/Getty Images; (c)Yann Arthus-Bertrand/CORBIS; (r)Jon Spaull/Getty Images; **640–641** Tyrone Turner/National Geographic Stock; **642** Patrick J. Endres/AlaskaPhotoGraphics.com; **643** Spencer Grant/PhotoEdit; **646** Creatas/SuperStock; **SR-00–SR-01** (bkgd)Gallo Images - Neil Overy/Getty Images; **SR-02** Hutchings Photography/Digital Light Source; **SR-06** Michell D. Bridwell/PhotoEdit; **SR-07** (t)The McGraw-Hill Companies; **SR-07** (b)Dominic Oldershaw; **SR-08** StudiOhio; **SR-09** Timothy Fuller; **SR-10** Aaron Haupt. **647** (t)Simon Fraser/Photo Researchers, Inc.; (b)Hutchings Photography/Digital Light Source; **650** Creatas/SuperStock; **651** George Diebold/Getty Images; **652** Unlisted Images/PhotoLibrary; **653** Russel Illig/Photodisc/Getty Images; **654** Clynt Garnham/Alamy; **658** (t)Clynt Garnham/Alamy; (c)Russel Illig/Photodisc/Getty Images; (b)Unlisted Images/PhotoLibrary; **660** Tyrone Turner/National Geographic Stock; **661** Ken Karp/digital light source; **664** Karen Huntt/Getty Images; **665** Jeff Greenberg/Alamy; **666** (t)Karen Huntt/Getty Images; (b)Jeff Greenberg/Alamy; **667** (t)Douglas McLaughlin; (b)Courtesy of Armour Homes, LLC; armourh.com; (bkgd)AFP/Getty Images; **668** James L. Amos/CORBIS; **669** Michelle D. Bridwell/PhotoEdit; **670** Klaus Nigge/Getty Images; **671** Hutchings Photography/Digital Light Source; **673** (t)Klaus Nigge/Getty Images; (b)James L. Amos/CORBIS; **675** Corbis/SuperStock; **676** (t)Patrick J. Endres/AlaskaPhotoGraphics.com; (b)Jeff Greenberg/Alamy; **679** Tyrone Turner/National Geographic Stock; **SR-00–SR-01** Gallo Images-Neil Overy/Getty Images; **SR-2** Hutchings Photography/Digital Light Source; **SR-6** Michell D. Bridwell/PhotoEdit; **SR-7** (t)The McGraw-Hill Companies, (b)Dominic Oldershaw; **SR-8** StudiOhio; **SR-9** Timothy Fuller; **SR-10** Aaron Haupt; **SR-12** KS Studios; **SR-13 SR-47** Matt Meadows; **SR-48** Stephen Durr, (c)NIBSC/Photo Researchers, Inc., (r)Science VU/Drs. D.T. John & T.B. Cole/Visuals Unlimited, Inc.; **SR-49** (t)Mark Steinmetz, (r)Andrew Syred/Science Photo Library/Photo Researchers, (br)Rich Brommer; **SR-50** David Fleetham/Visuals Unlimited/Getty Images, (l)Lynn Keddie/Photolibrary, (tr)G.R. Roberts; **SR-51** Gallo Images/CORBIS.

PERIODIC TABLE OF THE ELEMENTS

Legend:
- Element — Hydrogen
- Atomic number — 1
- Symbol — H
- Atomic mass — 1.01
- State of matter

- 🎈 Gas
- 💧 Liquid
- ⬜ Solid
- ⊙ Synthetic

*A column in the periodic table is called a **group**.*

*A row in the periodic table is called a **period**.*

The number in parentheses is the mass number of the longest lived isotope for that element.

Group	1	2	3	4	5	6	7	8	9
1	Hydrogen 1 H 1.01								
2	Lithium 3 Li 6.94	Beryllium 4 Be 9.01							
3	Sodium 11 Na 22.99	Magnesium 12 Mg 24.31							
4	Potassium 19 K 39.10	Calcium 20 Ca 40.08	Scandium 21 Sc 44.96	Titanium 22 Ti 47.87	Vanadium 23 V 50.94	Chromium 24 Cr 52.00	Manganese 25 Mn 54.94	Iron 26 Fe 55.85	Cobalt 27 Co 58.93
5	Rubidium 37 Rb 85.47	Strontium 38 Sr 87.62	Yttrium 39 Y 88.91	Zirconium 40 Zr 91.22	Niobium 41 Nb 92.91	Molybdenum 42 Mo 95.96	Technetium 43 Tc (98)	Ruthenium 44 Ru 101.07	Rhodium 45 Rh 102.91
6	Cesium 55 Cs 132.91	Barium 56 Ba 137.33	Lanthanum 57 La 138.91	Hafnium 72 Hf 178.49	Tantalum 73 Ta 180.95	Tungsten 74 W 183.84	Rhenium 75 Re 186.21	Osmium 76 Os 190.23	Iridium 77 Ir 192.22
7	Francium 87 Fr (223)	Radium 88 Ra (226)	Actinium 89 Ac (227)	Rutherfordium 104 Rf (267)	Dubnium 105 Db (268)	Seaborgium 106 Sg (271)	Bohrium 107 Bh (272)	Hassium 108 Hs (270)	Meitnerium 109 Mt (276)

Lanthanide series:

Cerium 58 Ce 140.12	Praseodymium 59 Pr 140.91	Neodymium 60 Nd 144.24	Promethium 61 Pm (145)	Samarium 62 Sm 150.36	Europium 63 Eu 151.96

Actinide series:

Thorium 90 Th 232.04	Protactinium 91 Pa 231.04	Uranium 92 U 238.03	Neptunium 93 Np (237)	Plutonium 94 Pu (244)	Americium 95 Am (243)